DFG Deutsche Forschungsgemeinschaft

Analyses of Hazardous Substances in Biological Materials

DFG Deutsche Forschungsgemeinschaft

Analyses of Hazardous Substances in Biological Materials

Volume 8

edited by Jürgen Angerer and Karl-Heinz Schaller
Working Group Analytical Chemistry

Commission for the Investigation of Health Hazards
of Chemical Compounds in the Work Area
(Chairman: Helmut Greim)

Deutsche Forschungsgemeinschaft
Kennedyallee 40, D-53175 Bonn, Federal Republic of Germany
Postal address: D-53175 Bonn
Phone: ++49/228/885-1
Telefax: ++49/228/885-2777
E-Mail: (Internet RFC 822): postmaster@dfg.de
Internet: http://www.dfg.de

Prof. Dr. Helmut Greim
Senatskommission
zur Prüfung gesundheitsschädlicher Arbeitsstoffe
der Deutschen Forschungsgemeinschaft
Technische Universität München
85350 Freising-Weihenstephan

Prof. Dr. Jürgen Angerer
Institut für Arbeits-, Sozial- und Umweltmedizin
Universität Erlangen-Nürnberg
Schillerstraße 25/29
91054 Erlangen

Translator: Elizabeth Mole

This book was carefully produced. Nevertheless, editors, authors and publisher do not warrant the information contained therein to be free of errors. Readers are advised to keep in mind that statements, data, illustrations, procedural details or other items may inadvertently be inaccurate.

Library of Congress Card No.: applied for

A catalogue record for this book is available from the British Library

Deutsche Bibliothek Cataloguing-in-Publication-Data:
Analyses of hazardous substances in biological materials /
DFG, Deutsche Forschungsgemeinschaft; Comm. for the Investigation
of Health Hazards of Chem. Compounds in the Work Area. –
Weinheim; New York; Chichester; Brisbane; Singapore; Toronto: Wiley-VCH
   ISSN 0179-7247
Vol. 1 (1985) –

ISBN 3-527-27791-9

© 2003 WILEY-VCH Verlag GmbH & Co. KgaA, Weinheim
Printed on acid-free paper.

All rights reserved (including those of translation in other languages). No part of this book may be reproduced in any form – by photoprinting, microfilm, or any other means – nor transmitted or translated into machine language without written permission from the publishers. Registered names, trademarks, etc. used in this book, even when not specifically marked as such, are not to be considered unprotected by law.

Composition: K+V Fotosatz GmbH, Beerfelden.
Printing: betz-druck gmbh, Darmstadt.
Bookbinding: Litges & Dopf Buchbinderei GmbH, Heppenheim.
Printed in the Federal Republic of Germany.

Preface

The 8th volume of Analyses of Hazardous Substances in Biological Materials contains 10 analytical methods to determine chemical substances and their metabolites in human body fluids. These methods principally serve to monitor compliance with the Biological Tolerance Values (BAT values) assigned by the Commission. Due to their high sensitivity they are suitable for checking reference values and human biomonitoring values (HBM values) that are established by the "Human Biomonitoring" Commission of the Umweltbundesamt [German Ministry for the Environment]. The methods permit an objective evaluation of the inner exposure to chemical substances in man in order to assess the health risk and to take appropriate precautionary measures if required. This volume broadens the spectrum of biological monitoring once again. The method for determination of 8-hydroxy-2-deoxyguanosine in urine deserves a special mention. This offers an investigation parameter which provides a specific and sensitive indication of the damage caused to human genetic material, and may permit conclusions on the individual risk of contracting cancer. The Commission deems it an important and urgent objective for the future activities of the "Analyses of Hazardous Substances in Biological Materials" working group to devise specific and sensitive methods to determine such "DNA adducts". Methods for the determination of polychlorinated dibenzodioxins and dibenzofurans as well as other persistent organochlorine compounds are included in this volume. In addition to the significance of these substances to environmental medicine, they are still of great importance to occupational medicine. Last but not least I would like to draw attention to the determination of cotinine in urine. With the help of this parameter it is possible to quantify the stress caused by passive smoking at the workplace, which has been evaluated as carcinogenic for humans by the Commission.

Germany has always been at the forefront of biological monitoring, and has thus led the field in endeavours to prevent health impairment to individuals as a result of exposure to chemicals. This volume must also be regarded as a contribution to the protection of health from the adverse influence of chemical substances in support of the international work in this field. Thanks are due to the Deutsche Forschungsgemeinschaft which has long sponsored this project. I would also like to extend my thanks to the chairman, Prof. Dr. Jürgen Angerer, as well as the members and guests of the „Analyses in Biological Materials" working group whose immense commitment has extended the possibilities for prevention and improved the diagnostic reliability of results in occupational and environmental medicine.

                        H. Greim
                        Chairman of the Deutsche Forschungsgemeinschaft's
                        Commission for the Investigation of Health Hazards
                        of Chemical Compounds in the Work Area

Foreword

The assay of toxic substances or their metabolites in human biological material enables determination of the dose of the toxin actually absorbed by the exposed person, and facilitates the estimation and if necessary the reduction of the resulting risk to health. Biological monitoring ("BM") has thus become such an indispensable tool for preventive healthcare in occupational medicine that it has been included in German and European occupational health and safety legislation. BM should be used more intensively in the field of environmental medicine to counteract the widespread tendency to imagine the worst-case scenario, and to return to a reasonable evaluation of actual environmental hazards.

The analytical methods published in this collection represent an essential basis for BM. These methods have been checked with regard to their analytical reliability and reproducibility. They are described in such detail that any analyst can immediately replicate them. At the same time, this collection provides information on those substances for which BM is possible and feasible to those involved in the fields of occupational and environmental medicine.

This volume extends the range of BM methods by a further 10 procedures, and also presents a general methodical chapter on the mineralisation of biological materials. The importance of GC/MS in occupational and environmental medicine today is clearly highlighted. In the meantime protein adducts of carcinogenic materials, such as benzylvaline, have become routine parameters for biochemical effect monitoring in well-equipped laboratories. Highly chlorinated, persistent organic substances, such as dioxines, PCB etc., are still an important issue in environmental medicine. In future BM of plant protection agents and cytostatic agents (isoproturon and oxazaphosphorines) will increasingly have to be taken into consideration. This certainly also applies to the contentious issue of "passive smoking at the workplace" (cotinine in urine). Moreover, the BM of DNA adducts will be of crucial significance in future. A first step in this direction has been taken by the inclusion of a specific method for the determination of 8-hydroxy-2-desoxiguanosine in this volume.

Foreword

I wish to take this opportunity to thank the members and guests of the "Analyses in Biological Materials" working group who have taken the progress of BM another step forward with the publication of this volume. Unreserved thanks are due to the Deutsche Forschungsgemeinschaft for its continuing support for the field of BM. An example has been set here that deserves to be emulated wherever man is exposed to chemical substances. It is important to me to express my thanks to Dr. Krawisch and Ms. Rosita Müller from the head office of the Deutsche Forschungsgemeinschaft as well as to Mr Tobias Weiß (Dipl. Chem.) and Ms. Gaidischki of the secretariat of the "Analyses in Biological Material" working group for their extremely constructive assistance and support.

                                    J. Angerer
                                    Chairman of the "Analytical Chemistry" working group
                                    of the Deutsche Forschungsgemeinschaft's
                                    Commission for the Investigation of Health Hazards
                                      of Chemical Compounds in the Work Area

Contents

Contents of Volumes 1–8 X

Working Group Analytical Chemistry of the Commission of the Deutsche Forschungsgemeinschaft for the Investigation of Health Hazards of Chemical Compounds in the Work Area XXVI

Organization – Objectives and operational procedure – Development, examination, release, and quality of the analytical methods – Publications of the working group – Withdrawal of methods

Terms and symbols used XXVIII

Terminology – symbols

Analytical methods

Digestion procedures for the determination of metals in biological materials	1
N-Benzylvaline after exposure to Benzylchloride in blood	35
Cotinine in urine	53
N,N-Dimethylacetamide (DMA) and N-methylacetamide (NMA) in urine .	67
Dioxins, furans and WHO PCB in whole blood	85
Hexamethylene diisocyanate (HDI) and hexamethylenediamine (HDA) in urine	119
8-Hydroxy-2'-deoxyguanosine in urine	133
1-(4-(1-Hydroxy-1-methylethyl)-phenyl)-3-methylurea (HMEPMU) as a metabolite of isoproturon in urine	151
Molybdenum in plasma and urine	167
Organochlorine compounds in whole blood and plasma	187
Oxazaphosphorines: Cyclophosphamide and ifosfamide in urine	221

Members and Guests of the Working Subgroup Analyses of Hazardous Substances in Biological Materials of the Commission of the Deutsche Forschungsgemeinschaft for the Investigation of Health Hazards of Chemical Compounds in the Work Area 239

Contents of Volumes 1–8

Substance	Vol.	Page
Acetone . . . . ,. . . .	see alcohols and ketones	
Acetylcholinesterase (AchE; acetylcholine-acetylhydrolase EC 3.1.1.7) in erythrocytes and cholinesterase (ChE: acylcholin-acylhydrolase EC 3.1.1.8) in plasma	3	45
Acetylcholine-acetylhydrolase	see acetylcholinesterase and cholinesterase	
AChE	see acetylcholinesterase and cholinesterase	
Acrylnitrile	see N-2-cyanoethylvaline, N-2-hydroxyethylvaline, N-methylvaline	
Acylcholin-acylhydrolase	see acetylcholinesterase and cholinesterase	
Alcohols and ketones (acetone; 1-butanol; 2-butanol; 2-butanone; ethanol; 2-hexanone; methanol; 2-methyl-1-propanol; 4-methyl-2-pentanone; 1-propanol, 2-propanol) in blood and urine	5	1
Aluminium, chromium, cobalt, copper, manganese, molybdenum, nickel, vanadium in urine	7	73
Aluminium in plasma	6	47
4-Aminodiphenyl	see aromatic amines	
4-Aminodiphenyl	see haemoglobin adducts of aromatic amines	
Aminotoluenes	see haemoglobin adducts of aromatic amines	
Amitrole (3-amino-1,2,4-triazole) in urine	6	63
Aniline	see aromatic amines	
Aniline	see haemoglobin adducts of aromatic amines	

Substance	Vol.	Page
o-anisidine	see aromatic amines	
Anthracycline cytostatic agents (doxorubicin, epirubicin, daunorubicin, idarubicin) in urine	7	119
Antimony in blood and urine	2	31
Antimony in urine	4	51
Antimony	see ICP-MS collective method	
Application of the ICP-MS for biological monitoring	6	1
Aromatic alcohols	see phenols and aromatic alcohols	
Aromatic amines in urine (1-naphthylamine; 2-naphthylamine; 4,4′-methylene-bis(2-chloroaniline); 3,3′-dichlorobenzidine)	1	17
Aromatic amines (aniline; *o*-toluidine; *m*-toluidine; *p*-toluidine; 2,4- and 2,6-toluylenediamine; 4-aminodiphenyl; 4,4′-diaminodiphenylmethane) in urine, plasma and erythrocytes	4	67
Aromatic carboxylic acids in urine (phenylglyoxylic acid; mandelic acid; hippuric acid; *o*-methylhippuric acid, *m*-/*p*-methylhippuric acids; benzoic acid)	2	47
Arsenic in urine	3	63
Arsenic Species (As(III), As(V), monomethylarsonic acid, dimethylarsinic acid in urine	7	97
As(III)	see arsenic species	
As(V)	see arsenic species	
Barium in urine	3	81
Barium, strontium, titanium in urine	2	67
Benzene	see benzene and alkylbenzenes	

Substance	Vol.	Page
Benzene	see *t,t*-muconic acid	
Benzene	see S-phenylmercapturic acid	
Benzene and alkylbenzenes (BTX-aromatics) in blood	4	107
Benzene derivatives in urine, suitable for steam distillation (phenol; *m-/p*-cresol; *o*-cresol; *o*-chlorophenol; *o*-nitrophenol; *p*-chlorophenol; nitrobenzene; 1,2-dinitrobenzene; 2-chloro-5-methylphenol; 2,5-dichlorophenol; 3,4-dichlorophenol; 2,3-dinitrotoluene)	1	31
Benzidine	see haemoglobin adducts of aromatic amines	
Benzoic acid	see aromatic carboxylic acids	
Benzylchloride	see N-Benzylvaline	
N-Benzylvaline after exposure to benzylchloride in blood	8	35
Beryllium in urine	1	57
Beryllium, lithium, vanadium, tungsten in urine	5	51
Beryllium in urine, standard addition procedure	5	35
Bismuth	see ICP-MS collective method	
Bromide in urine	1	67
2-Bromo-2-chloro-1,1,1-trifluoroethane	see halogenated hydrocarbons	
1-Butanol	see alcohols and ketones	
2-Butanol	see alcohols and ketones	
2-Butanone	see alcohols and ketones	
Butoxyacetic acid in urine	4	131
Cadmium in blood	1	79
Cadmium in urine	2	85

Substance	Vol.	Page
Cadmium	see ICP-MS collective method	
Carboxyhemoglobin in blood	1	93
ChE	see acetylcholinesterase and cholinesterase	
Chlorinated aromatic hydrocarbons in plasma (o-dichlorobenzene; m-dichlorobenzene; p-dichlorobenzene; 1,2,4-trichlorobenzene; 1,2,4,5-tetrachlorobenzene; pentachlorobenzene; hexachlorobenzene)	3	93
4-chloroaniline	see haemoglobin adducts of aromatic amines	
p-chloroaniline	see haemoglobin adducts of aromatic amines	
Chlorobenzenes in blood (1,2- and 1,4-dichlorobenzene; 1,2,4-trichlorobenzene)	1	107
2-Chloro-5-methylphenol	see benzene derivatives	
4-Chloro-2-methylphenoxyacetic acid	see chlorophenoxycarboxylic acids	
4-Chloro-2-methylphenoxypropionic acid	see chlorophenoxycarboxylic acids	
o-Chlorophenol, p-chlorophenol	see benzene derivatives	
Chlorophenols in urine (2,4-dichlorophenol, 2,5-dichlorophenol, 2,6-dichlorophenol, 2,3,4-trichlorophenol, 2,4,5-trichlorophenol, 2,4,6-trichlorophenol, 2,3,4,6-tetrachlorophenol)	7	143
4-Chloro-o-toluidine	see aromatic amines	
Chlorophenols (monohydroxychlorobenzenes) in urine (2,6-, 2,3-, 3,4-dichlorophenol; 2,4,6-, 2,4,5-, 3,4,5-trichlorophenol; 2,3,4,6-, 2,3,4,5-tetrachlorophenol; pentachlorophenol) ..	1	123

Substance	Vol.	Page
Chlorophenoxycarboxylic acids (4-chloro-2-methylphenoxyacetic acid; 2,4-dichlorophenoxyacetic acid; 4-chloro-2-methylphenoxypropionic acid; 2,4-dichlorophenoxypropionic acid) in urine	5	77
Cholinesterase	see acetylcholinesterase and Cholinesterase	
Chromium	see aluminium, chromium, cobalt, copper, manganese, molybdenum, nickel, vanadium	
Chromium in urine	2	97
Chromium in whole blood, plasma and erythrocytes	3	109
Cobalt	see aluminium, chromium, cobalt, copper, manganese, molybdenum, nickel, vanadium	
Cobalt in urine	1	141
Cobalt in blood	2	117
Copper	see aluminium, chromium, cobalt, copper, manganese, molybdenum, nickel, vanadium	
Cotinine in urine, plasma or serum	7	171
Cotinine in urine	8	53
o-Cresol; m-/p-cresol	see benzene derivatives	
Cyanide in blood	2	133
N-2-Cyanoethylvaline, N-2-hydroxyethylvaline, N-methylvaline in blood	5	211
Cyclophosphamide	see Oxazaphosphorines	

Contents of Volumes 1–8

Substance	Vol.	Page
Daunorubicin	see anthracycline cytostatic agents	
DDE	see Oxachlorine compounds	
DDT	see Oxachlorine compounds	
2,4-Diamino-6-chloro-s-triazine in urine	6	111
4,4'-Diaminodiphenylmethane	see aromatic amines	
4,4'-Diaminodiphenylmethane	see haemoglobin adducts of aromatic amines	
Dibenzodioxins	see Dioxins, furans and WHO PCB	
Dibenzofurans	see Dioxins, furans and WHO PCB	
cis-3-(2,2-dibromovinyl)-2,2-dimethylcyclopropane-1-carboxylic acid	see pyrethroid metabolites	
1,2-; 1,4-Dichlorobenzene	see chlorobenzenes	
o-dichlorobenzene, m-dichlorobenzene, p-dichlorobenzene	see chlorinated aromatic hydrocarbons	
3,3'-Dichlorobenzidine	see aromatic amines	
3,3'-Dichlorobenzidine	see haemoglobin adducts of aromatic amines	
4,4'-Dichlorodiphenyldichloroethane	see Oxachlorine compounds	
4,4'-Dichlorodiphenyltrichloroethane	see Oxachlorine compounds	
1,2-Dichloroethylene	see chlorinated aromatic hydrocarbons	
Dichloromethane	see chlorinated aromatic hydrocarbons	
2,5-; 3,4-Dichlorophenol	see benzene derivatives	
2,6-; 2,3-; 3,4-Dichlorophenol	see chlorophenols	

Substance	Vol.	Page
2,4-Dichlorophenol	see chlorophenols	
2,5-Dichlorophenol	see chlorophenols	
2,6-Dichlorophenol	see chlorophenols	
2,4-Dichlorophenoxyacetic acid	see chlorophenoxycarboxylic acids	
2,4-Dichlorophenoxypropionic acid	see chlorophenoxycarboxylic acids	
cis-3-(2,2-dichlorovinyl)-2,2-dimethylcyclopropane-1-carboxylic acid	see pyrethroid metabolites	
trans-3-(2,2-dichlorovinyl)-2,2-dimethylcyclopropane-1-carboxylic acid	see pyrethroid metabolites	
Digestion procedures for the determination of metals in biological material	2	1
Digestion procedures for the determination of metals in biological materials	8	1
3,4-Dihydroxychlorobenzene (4-chlorocatechol) in urine	6	125
N,N-Dimethylacetamide (DMA) and N-methylacetamide (NMA) in urine	8	67
Dimethylarsinic acid	see arsenic species	
N,N-Dimethylformamide (DMF) in urine	5	97
2,4-; 2,3-; 3,4-Dimethylphenol	see phenols and aromatic alcohols	
1,2-Dinitrobenzene	see benzene derivatives	
o-dinitrobenzene	see nitroaromatic compounds	
2,3-Dinitrotoluene	see benzene derivatives	
2,6-Dinitrotoluene	see nitroaromatic compounds	
Dioxins, furans and WHO PCB in whole blood	8	85

Substance	Vol.	Page
Doxorubicin	see anthracycline cytostatic agents	
Epirubicin	see anthracycline cytostatic agents	
Erythrocyte porphyrins (free) in blood (erythrocytes)	2	145
Ethanol	see alcohols and ketones	
Ethylbenzene	see benzene and alkyl benzenes	
Ethylene oxide	see N-2-cyanoethylvaline, N-2-hydroxyethylvaline, N-methylvaline	
2-Ethylphenol	see phenols and aromatic alcohols	
Fluoride in urine	2	159
4-Fluoro-3-phenoxybenzoic acid	see pyrethroid metabolites	
Free erythrocyte porphyrins	see erythrocyte porphyrins (free)	
Furans	see Dioxins, furans and WHO PCB	
Gas chromatographic methods for the determination of organic substances in biological material	3	1
Gold	see platinum and gold	
Haemoglobin adducts of aromatic amines (aniline, *o*-, *m*- and *p*-toluidine, *o*-anisidine, *p*-chloroaniline, α- and β-naphthylamine, 4-aminodiphenyl, benzidine, 4,4'-diaminodiphenylmethane, 3,3'-dichlorobenzidine)	7	191
Halogenated hydrocarbons in blood (dichloromethane; 1,2-dichloroethylene; 2-bromo-2-chloro-1,1,1-trifluoroethane (halothane); trichloromethane; 1,1,1-trichloroethane; tetrachloromethane; trichloroethylene; tetrachloroethylene)	3	127

Substance	Vol.	Page
Halothane	see halogenated hydrocarbons	
Hexachlorobenzene	see chlorinated aromatic hydrocarbons and oxachlorine compounds	
Hexachlorocyclohexane	see Oxachlorine compounds	
Hexamethylenediamine	see Hexamethylene diisocyanate (HDI) and hexamethylenediamine	
Hexamethylene diisocyanate (HDI) and hexamethylenediamine (HDA) in urine	8	119
2,5-Hexanedione	see hexane metabolites	
Hexane metabolites (2,5-hexanedione, 2-hexanone) in urine	4	147
2-Hexanone	see hexane metabolites	
2-Hexanone	see alcohols and ketones	
Hippuric acid	see aromatic carboxylic acids and o-, m-/p-Methylhippuric acids	
Hydrazine in blood (plasma)	2	171
Hydrazine and **N-acetylhydrazine** in urine and plasma	6	141
8-Hydroxy-2′-deoxyguanosine in urine	8	133
1-(4-(1-Hydroxy-1-methylethyl)-phenyl)-3-methylurea (HMEPMU)	8	151
1-Hydroxyphenanthrene	see PAH metabolites	
4-Hydroxyphenanthrene	see PAH metabolites	
9-Hydroxyphenanthrene	see PAH metabolites	
1-Hydroxypyrene in urine	3	151
1-Hydroxypyrene in urine	see PAH metabolites	

Substance	Vol.	Page
ICP-MS collective method (antimony, bismuth, cadmium, lead, mercury, platinum, tellurium, thallium, tin, tungsten) in urine	6	79
Idarubicin	see anthracycline cytostatic agents	
Ifosfamide	see Oxazaphosphorines	
Indium in urine	3	171
Isoproturon	see 1-(4-(1-Hydroxy-1-methylethyl)-phenyl)-3-methylurea (HMEPMU)	
Lead in blood	1	155
Lead in blood and urine	2	183
Lead	see ICP-MS collective method	
Lithium	see beryllium, lithium, vanadium, tungsten	
Mandelic acid	see aromatic carboxylic acids	
Manganese	see aluminium, chromium, cobalt, copper, manganese, molybdenum, nickel, vanadium	
Mercury in blood and urine	2	195
Mercury	see ICP-MS collective method	
Metasystox R in urine	7	221
Methanol	see alcohols and ketones	
N-Methylacetamide	see N,N-dimethylacetamide (DMA) and N-methylacetamide (NMA)	
Methylating agents (bis(chloromethyl)ether, bromomethane, chloromethane, dimethyl sulphate, iodomethane, monochlorodimethyl ether)	see N-2-cyanoethylvaline, N-2-hydroxyethylvaline, N-2-methylvaline	

Contents of Volumes 1–8

Substance	Vol.	Page
3-Methylbenzyl alcohol	see phenols and aromatic alcohols	
4,4′-Methylene-*bis*(2-chloroaniline)	see aromatic amines	
o-Methylhippuric acid, m-/p-methylhippuric acids (Toluric acids), hippuric acid in urine	1	165
o-Methylhippuric acid, m-/p-methylhippuric acids	see aromatic carboxylic acids	
4-Methyl-2-pentanone	see alcohols and ketones	
2-, 4-Methylphenol	see phenols and aromatic alcohols	
2-Methyl-1-propanol	see alcohols and ketones	
β_2-Microglobulin in urine and serum	3	185
Molybdenum	see aluminium, chromium, cobalt, copper, manganese, molybdenum, nickel, vanadium	
Molybdenum in urine	5	109
Molybdenum in plasma and urine	8	167
Monohydroxychlorobenzenes	see chlorophenols	
Monomethylarsonic acid	see arsenic species	
t,t-Muconic acid in urine	5	125
1-, 2-Naphthylamine	see aromatic amines	
1-Naphthylamine	see haemoglobin adducts of aromatic amines	
2-Naphthylamine	see haemoglobin adducts of aromatic amines	
α-Naphthylamine	see haemoglobin adducts of aromatic amines	
β-Naphthylaminex	see haemoglobin adducts of aromatic amines	

Substance	Vol.	Page
Nickel	see aluminium, chromium, cobalt, copper, manganese, molybdenum, nickel, vanadium	
Nickel in blood	3	193
Nickel in urine	1	177
Nitroaromatic compounds in plasma (nitrobenzene; *p*-nitrotoluene, *p*-nitrochlorobenzene; 2,6-dinitrotoluene; *o*-dinitrobenzene; 1-nitronaphthalene; 2-nitronaphthalene; 4-nitrobiphenyl)	3	207
Nitrobenzene	see benzene derivatives	
Nitrobenzene	see nitroaromatic compounds	
4-Nitrobiphenyl	see nitroaromatic compounds	
p-Nitrochlorobenzene	see nitroaromatic compounds	
1-, 2-Nitronaphthaline	see nitroaromatic compounds	
o-Nitrophenol	see benzene derivatives	
p-Nitrotoluene	see nitroaromatic compounds	
Organochlorine compounds in whole blood and plasma	8	187
Organotin compounds (except methyltin compounds) and total tin in urine	4	165
Oxazaphosphorines: Cyclophosphamide and ifosfamide in urine	8	221
Oxydemeton-methyl in urine	see metasystox R	
PAH metabolites in urine	6	163
Pentachlorobenzene	see chlorinated aromatic hydrocarbons	
Pentachlorophenol	see chlorophenols	
Pentachlorophenol in urine	7	237

Substance	Vol.	Page
Pentachlorophenol in urine and serum/plasma	6	189
Phenol in urine	1	189
Phenol	see benzene derivatives and phenols and aromatic alcohols	
Phenols and aromatic alcohols in urine (phenol; 2- and 4-methylphenol; DL-1- and 2-phenylethanol; 3-methylbenzyl alcohol; 2-ethylphenol; 2,4-; 2,3- and 3,4-dimethylphenol	2	213
Phenols in urine	6	211
3-Phenoxybenzoic acid	see pyrethroid metabolites	
DL-1-, 2-phenylethanol	see phenols and aromatic alcohols	
Phenylglyoxylic acid	see aromatic carboxylic acids	
S-Phenylmercapturic acid in urine	5	143
Platinum in urine, blood, plasma/serum	4	187
Platinum	see ICP-MS collective method	
Platinum and gold in urine	7	255
Polychlorinated biphenyls in blood or serum	3	231
Polychlorinated biphenyls	see Dioxins, furans and WHO PCB and see Oxachlorine compounds	
Polychlorinated dibenzodioxines	see Dioxins, furans and WHO PCB	
Polychlorinated dibenzofurans	see Dioxins, furans and WHO PCB	
1-Propanol	see alcohols and ketones	
2-Propanol	see alcohols and ketones	

Substance	Vol.	Page
Pyrethroid metabolites (*cis*-3-(2,2-dichlorovinyl)-2,2-dimethylcyclopropane-1-carboxylic acid, *trans*-3-(2,2-dichlorovinyl)-2,2-dimethylcyclopropane-1-carboxylic acid, *cis*-3-(2,2-dibromovinyl)-2,2-dimethylcyclopropane-1-carboxylic acid, 4-fluoro-3-phenoxybenzoic acid, 3-phenoxybenzoic acid) in urine	6	231
Rhodium in urine and serum/plasma	7	273
Selenium in blood, plasma and urine	2	231
Strontium	see barium, strontium, titanium	
Tellurium	see ICP-MS collective method	
1,2,4,5-Tetrachlorobenzene	see chlorinated aromatic hydrocarbons	
Tetrachloroethylene	see halogenated hydrocarbons	
Tetrachloromethane	see halogenated hydrocarbons	
2,3,4,6-; 2,3,4,5-Tetrachlorophenol	see chlorophenols	
Thallium in urine	1	199
Thallium in urine	5	163
Thallium	see ICP-MS collective method	
The use of atomic absorption spectrometry for the determination of metals in biological materials	4	1
2-Thioxothiazoldine-4-carboxylic acid (TTCA) in urine	4	207
Thorium	see thorium and uranium	
Thorium and **uranium** in urine	6	255
Tin in urine	4	223
Tin	see ICP-MS collective method	
Titanium	see barium; strontium, titanium	

Substance	Vol.	Page
Toluene	see benzene and alkylbenzenes	
Toluidines	see haemoglobin adducts of aromatic amines	
m-Toluidines	see haemoglobin adducts of aromatic amines	
o-Toluidines	see haemoglobin adducts of aromatic amines	
p-Toluidines	see haemoglobin adducts of aromatic amines	
o-Toluidine, *m*-Toluidine, *p*-Toluidine	see aromatic amines	
Toluric acids	see methylhippuric acids	
2,4- and 2,6-Toluylendiamine	see aromatic amines	
Trichloroacetic acid (TCA) in urine	1	209
1,2,4-Trichlorobenzene	see chlorobenzenes and chlorinated aromatic hydrocarbons	
Trichloroethane	see halogenated hydrocarbons	
Trichloroethylene	see halogenated hydrocarbons	
Trichloromethane	see halogenated hydrocarbons	
2,4,6-; 2,4,5-; 3,4,5-Trichlorophenol	see chlorophenols	
2,3,4-Trichlorophenol	see chlorophenol	
2,4,5-Trichlorophenol	see chlorophenol	
2,4,6-Trichlorophenol	see chlorophenol	
Tungsten	see beryllium, lithium, vanadium, tungsten and ICP-MS collective method	
Uranium	see thorium and uranium	
Vanadium	see aluminium, chromium, cobalt, copper, manganese, molybdenum, nickel, vanadium	

Substance	Vol.	Page
Vanadium in urine	3	241
Vanadium	see beryllium, lithium, vanadium, tungsten	
Vinclozolin as 3,5-dichloroaniline in urine	7	287
3,5-Dichloroaniline	see vinclozolin	
o-xylene, *m*-xylene	see benzene and alkylbenzenes	
Zinc in plasma, serum and urine	5	211

Working Group Analytical Chemistry of the Commission of the Deutsche Forschungsgemeinschaft for the Investigation of Health Hazards of Chemical Compounds in the Work Area

Organization

The Working Group Analytical Chemistry (Chairman: J. Angerer) was established in 1969. It includes two Working Subgroups: Air Analyses (Leader: A. Kettrup), Analyses of Hazardous Substances in Biological Materials (Leaders: J. Angerer, K. H. Schaller). The participants, who have been invited to collaborate on a Working Subgroup by the leaders, are experts in the field of technical and medical protection against chemical hazards at the workplace.

A list of the members and guest of Analyses of Hazardous Substances in Biological Materials is given at the end of this volume.

Objectives and operational procedure

The two analytical subgroups are charged with the task of preparing methods for the determination of hazardous industrial materials in the air of the work place or to determine these hazardous materials or their metabolic products in biological specimens from the persons working there. Within the framework of the existing laws and regulations, these analytical methods are useful for ambient monitoring at the work place and biological monitoring of the exposed persons.

In addition to working out the analytical procedure, these subgroups are concerned with the problems of the preanalytical phase (specimen collection, storage, transport), the statistical quality control, as well as the interpretation of the results.

Development, examination, release, and quality of the analytical methods

In its selection of suitable analytical methods, the Working Group is guided mainly by the relevant scientific literature and the expertise of the members and guests of the Working Subgroup. If appropriate analytical methods are not available they are worked out within the Working Group. The leader designates an author, who assumes the task of developing and formulating a method proposal. The proposal is examined experimentally by at least one other member of the project, who then submits a written report of the results of the examination. As a matter of principle the examination must encompass all phases of the proposed analytical procedure.

The examined method is then laid before the members of the subgroup for consideration. After hearing the judgement of the author and the examiner they can approve the method. The method can then be released for publication after a final meeting of the leader of the Working Group Analytical Chemistry with the subgroup leaders, authors, and examiners of the method.

Under special circumstances an examined method can be released for publication by the leader of the Working Group after consultation with the subgroup leaders.

Only methods for which criteria of analytical reliability can be explicitly assigned are released for publication. The values for inaccuracy, imprecision, detection limits, sensitivity, and specificity must fulfill the requirements of statistical quality control as well as the specific standards set by occupational health. The above procedure is meant to guarantee that only reliably functioning methods are published, which are not only reproducible within the framework of the given reliability criteria in different laboratories but also can be monitored over the course of time.

In the selection and development of a method for determining a particular substance the Working Group has given the analytical reliability of the method precedence over aspects of simplicity and economy.

Publications of the working group

Methods released by the Working Group are published in the Federal Republic of Germany by the Deutsche Forschungsgemeinschaft as a loose-leaf collection entitled "Analytische Methoden zur Prüfung gesundheitsschädlicher Arbeitsstoffe" (WILEY-VCH, Weinheim, FRG). The collection at present consists of two volumes:
Volume I: Luftanalysen
Volume II: Analysen in biologischem Material.
These methods are published in an English edition. The work at hand represents the sixth English issue of "Analyses of Hazardous Substances in Biological Materials".

Withdrawal of methods

An analytical method that is made obsolete by new developments or discoveries in the fields of instrumental analysis or occupational health and toxicology can be replaced by a more efficient method. After consultation with the membership of the relevant project and with the consent of the leader of the Working Group, the subgroup leader is empowered to withdraw the old method.

Terms and symbols used

Terminology

Accuracy – the agreement between the best estimate of a quantity and its true value. It has no numerical value.
Analyte – the component to be measured.
Assigned value – the concentration of the analyte in the control specimen assigned either arbitrarily, e.g., convention, or from preliminary evidence, e.g., in the absence of a recognized reference method.

Biological Tolerance Value for a Working Material (BAT) – is defined as the maximum permissible quantity of a chemical compound, its metabolites, or any deviation from the norm of biological parameters induced by these substances in exposed humans. According to current knowledge these conditions generally do not impair the health of the employee, even if exposure is repeated and of long duration. As with MAK values, the maximum period of exposure to a working material is generally given as eight hours daily and 40 hours weekly.
Blank value – the analytical result obtained when the complete procedure is carried out on ultrapure water containing no analyte instead of biological specimens.

Calibration standards – specimens with known concentrations of the analyte that are used for calibration.
Certified value – the concentration of the analyte in control specimens certified by an official body subject to conditions established by that body.
Control material – a material that is used solely for quality control purposes and not for calibration.

Definitive method – the analytical method of all the methods for determining the analyte that is capable of providing the highest accuracy. Its accuracy must be adequate for its stated purposes.
Definitive value – the concentration of analyte in control specimens determined by a definitive method. It is the best available estimate of the true value.
Detection limit – the minimum analytical result that is still clearly detectable and distinguishable from the background noise; defined as three standard deviations of the appropriate blank value.

External quality control – a procedure of utilizing for quality control purposes the results of analyses performed on the same specimen or specimens by several laboratories.

Imprecision – the standard deviation or coefficient of variation of the results in a set of replicate measurements. A distinction is made between within-series, between-day, and interlaboratory imprecision.
Inaccuracy – the numerical difference between the mean of a set of replicate measurements and the true value.

Influence factors – lead to changes in vivo in the clinical chemical parameter. Their influences are independent of the specificity of the analytical method.
Interference – the effect on the accuracy of measurement of one component caused by another component that does not itself produce a reading.
Interference factors – all factors that alter the result in vitro, i.e., after the specimen has been collected from the patient.
Internal quality control – the procedure of utilizing the results of only one laboratory for quality control purposes. It includes the control of imprecision as well as inaccuracy.

Maximum Concentration Value at the Workplace (MAK) – is defined as the maximum permissible concentration of a chemical compound present in the air within a working area (as gas, vapor particulate matter) which, according to current knowledge, generally does not impair the health of the employee nor cause undue annoyance. Under these conditions, exposure can be repeated and of long duration over a daily period of eight hours, constituting an average work week of 40 hours (42 hours per week as averaged over four successive weeks for firms having four work shifts).
Measure – the measured change of one parameter of the analyte in the physical or chemical system used for analysis.

Preanalytical phase – the period from the specimen collection to the aliquotation (sampling) of the biological specimens.
Precision – the agreement between replicate measurements. It has no numerical value.
Prognostic range – an interval that with a given probability includes the analytical result from an identical specimen.

Quality control – the study of those errors that are the responsibility of the laboratory and the procedures used to recognize and minimize them. This study includes all errors arising within the laboratory during the time from aliquotation of the specimen to dispatch of the report.

Recovery rate – in recovery experiments the amount of recovered analyte divided by the amount of added analyte expressed as a percentage.
Reference material – a material or substance for which one or more properties are sufficiently well established to be used for calibration of an apparatus or for verification of an analytical method.
Reference method – an analytical method whose inaccuracy and imprecision are small enough as demonstrated by direct comparison with the definitive method and show low incidence of susceptibility to known interferences is so well documented that the stated aims of the reference method may be achieved.
Reference method value – the most probable value derived from a set of results obtained by the most reliable reference methods available.
Reliability criteria – defined quantifiable parameters for the assessment of the quality of an analytical method, e.g., imprecision, inaccuracy, detection limit.

Terms and symbols used

Sample – that appropriate representation portion of a specimen used in the analysis.
Sensitivity – the differential quotient of the calibration function.
Selected method – routine method with known systematic error.
Specificity – the ability of an analytical method to determine solely the component or components it purports to measure. It has no numerical value.
Specimen – the material available for analysis.

Symbols

c	substance concentration of analyte
$\bar{c}$	mean substance concentration of analyte
E	extinction
k'	reciprocal calibration factor
m	mass
M	molar mass
n	number
p	pressure
P	probability
r	recovery rate
s	relative standard deviation
$s_{bl,abs}$	standard deviation of the blank value
s_d	relative standard deviation derived from duplicate analyses
s_w	relative standard deviation derived from replicate analyses of the same specimen
t_p	Student's t factor
T	thermodynamic temperature
u	prognostic range
V	volume
V_m	molar volume
x	observed measure of analyte
$\bar{x}$	mean
x_{bl}	observed measure of blank
z	number of duplicate analyses
H	sensitivity
ϱ	mass concentration
σ	volume concentration

Digestion procedures for the determination of metals in biological materials

Contents

1   Introduction
2   Available digestion methods
3   Vessel materials for digestion reactions
3.1   Vessel materials for wet digestion
3.1.1   Quartz
3.1.2   Borosilicate glass
3.1.3   Polytetrafluoroethylene (PTFE)
3.1.4   Perfluoroalkoxy polymer (PFA)
3.1.5   Glassy carbon
4   Digestion reagents
4.1   Oxidising agents
4.1.1   Nitric acid
4.1.2   Hydrogen peroxide
4.1.3   Potassium permanganate
4.1.4   Perchloric acid
4.2   Solubilisers and neutralising agents
4.2.1   Sulphuric acid
4.2.2   Hydrochloric acid
4.2.3   Ammonia
5   Digestion techniques
5.1   Combustion with oxygen in a closed system
5.1.1   Combustion in autoclaves
5.1.2   Schöniger combustion to determine iodine in biological material
5.2   Wet digestion
5.2.1   Open systems
5.2.2   Closed systems
6   Current "state-of-the-art" for digestion procedures
7   Appendix
7.1   Examples of procedures

1 Introduction

7.1.1 Example 1: Digestion under oxygen in a closed system
7.1.2 Example 2: Schöniger combustion for the determination of iodine
7.1.3 Example 3: Open ashing with HNO_3/H_2O_2 (Fenton's reagent)
7.1.4 Example 4: Open digestion with HNO_3/H_2SO_4
7.1.5 Example 5: Open digestion with potassium permanganate for the determination of Hg
7.1.6 Example 6: Digestion under low overpressure (<5 bar)
7.1.7 Example 7: Digestion under medium overpressures (<20 bar) using a "Seif apparatus"
7.1.8 Example 8: High-pressure digestion with microwave heating (e.g. pressurised microwave decomposition (PMD))
7.1.9 Example 9: High-pressure digestion in the "Multiwave"
7.1.10 Example 10: High-pressure digestion using the "High Pressure Asher (HPA)"
7.1.11 Example 11: Digestion of liquids in a high-pressure microwave flowthrough digestion apparatus
7.1.12 Example 12: UV digestion of liquids
8 References

1 Introduction

Most methods for the determination of trace elements in biological materials require digestion of the matrix (destruction of the organic matrix by means of oxidation).
There are numerous reasons for destroying and dissolving the organic matrix:
Dissolution of solid samples (tissue, bones, etc.) is necessary, as almost all the common analytical procedures are suitable for analysing solutions only. In the case of the analysis of blood or urine, previous digestion is required to enable intermediate chemical steps, such as hydride formation, enrichment procedures, etc. to be carried out.
Separation of interfering components of the sample, especially of the matrix, is necessary because of possible chemical or physical matrix effects.
Enrichment of elements is required if the detection capability of the selected analytical method is not sufficient. This can only be achieved after destruction of the organic matrix.
Isoformation of the sample and standard solution (in particular the type and concentration of the acid used for decomposition in this case) prevents the possible occurrence of changes in sensitivity.
Finally, digestion leads to a certain degree of *homogenisation* of the material compared with direct analysis of the sample.
The selection of the digestion method largely depends on three criteria: the nature of the matrix, the element to be determined and the analytical method to be used.
An optimum digestion method should meet the following requirements:
- The biological matrix must be completely destroyed.
- The content of the elements to be determined may not be changed (neither by loss nor by contamination).

- After digestion, the elements to be analysed must be in homogeneous solutions and in a suitable form for the subsequent determination.

2 Available digestion methods

Due to the diversity of sample material, the elements to be determined and the analytical methods, no universally valid procedure can be recommended. On the one hand, tried-and-tested and newly developed digestion methods will be described for certain problems, and on the other, information and general considerations for devising new digestion methods will be presented in the following sections.
Finally, practical examples are given in the appendix.
On principle, the following digestion methods can be distinguished and used:
1. Digestion by melting
   - Basic
   - Acidic
2. Combustion with oxygen
   - In open systems
     - Dry ashing
     - Ashing at low temperatures
     - Hydrogen-oxygen flame ("Wickbold")
   - In closed systems
     - Flask combustion ("Schöniger")
     - Oxygen "bombs"
3. Wet ashing
   - In open systems
   - In closed systems ("pressurised ashing")
   - In flowthrough systems.

The procedures involving *digestion by melting* are of practically no importance for the analysis of organic material.
Combustion with oxygen plays rather a minor role in practice, but it is very effective in certain cases and will therefore be briefly described here.
Dry ashing techniques in a muffle oven, which are still widely used today – especially for routine analysis – should be avoided for trace analysis on principle, as considerable losses of elements occur (even at low temperatures, see also Table 1), and also because of the danger of contamination (systematic errors). Therefore these techniques will not be discussed here.
Low-temperature ashing using oxygen plasmas is effective, especially in a closed system, but is very time-consuming in practice. At present no suitable equipment is commercially available for this technique.
Although *combustion* of organic samples (solid and liquid) in an *oxygen-hydrogen flame* (Wickbold combustion) is very suitable for the determination of anions, it only has drawbacks for trace elements (large quantities of solvent, glass apparatus with a large surface, incomplete combustion in some cases).

Table 1. Element losses during dry ashing (excerpt from [1]) (0: no loss found; +: 2 to 5% loss; ++: 6 to 20% loss; +++: >20% loss; n.d. = no determination carried out).

Element	400 °C	450 °C	500 °C	550 °C	600 °C	700 °C
As	+++	+++	+++	+++	n.d.	+++
Cd	0	0	++	n.d.	n.d.	n.d.
Co	+++	+++	+++	n.d.	+++	n.d.
Cr	0	0	n.d.	n.d.	+++	++
Cu	0	+++	++	+++	n.d.	++
Hg	+++	n.d.	+++	+++	n.d.	+++
Pb	0	+	+++	n.d.	n.d.	+++
Sb	+++	n.d.	n.d.	n.d.	n.d.	+++
Zn	0	++	+++	+++	n.d.	+++

Wet ashing is certainly the most widely used technique. The procedure varies, depending on whether it is carried out in an open or a closed system, and on the type of heat supply (conventional or microwave heating).

3 Vessel materials for digestion reactions

Most digestion techniques are carried out under extreme conditions as regards the temperature and/or the digestion reagents used. Therefore the material of the vessel must be carefully selected, taking the digestion method into consideration. The following criteria must be borne in mind when evaluating the suitability of materials:
Heat-resistance, thermal conductivity, mechanical strength, resistance to acids or alkalis, surface properties, formation of compounds or contamination, whereby it is particularly important to consider the special characteristics of the biological material and the generally very low analyte concentrations. Digestion vessels can be made of platinum and platinum alloys, gold, silver, nickel, stainless steel, tantalum, porcelain, **borosilicate glass, quartz, PTFE** (polytetrafluoroethylene), **PFA** (perfluoroalkoxy polymer: copolymer of tetrafluoroethylene and perfluorinated co-components) and glassy carbon, whereby the four latter materials are the most commonly used for biological samples.
In particular, the various metals and porcelain are unsuitable for modern trace element analysis due to the considerable risk of contamination, and they will therefore not be discussed here.
Possible contamination must always be considered and checked in each individual case and depending on the task in question. Table 2 shows the typically occurring elemental contamination for some of the vessel materials described below.

3 Vessel materials for digestion reactions

Table 2. Examples of typically occurring contamination of elements (in ng/L) in digestion vessels. The values given here are not necessarily representative.

Element	Glassy carbon Sigradur G®	PTFE Teflon	Quartz Heralux	Quartz Suprasil®	Borosilicate glass
B	100	–	100	10	main component
Na	300	25000	1000	10	main component
Mg	100	–	100	100	6×10^5
Al	6000	–	30000	100	main component
Si	85000	–	main component	main component	main component
Ca	80000	–	800–3000	100	106
Ti	12000	–	800	100	3000
Cr	80	30	5	3	3000
Mn	100	–	10	10	6000
Fe	2000	10	800	200	2×10^5
Co	2	2	1	1	100
Ni	500	–	–	–	2000
Cu	200	20	–	10	1000
Zn	300	10	50	100	3000
As	50	–	80	0.1	500–22000
Cd	10	–	10	>	1000
Sb	10	0.4	2	1	8000
Hg	1	10	1	1	–

3.1 Vessel materials for wet digestion

3.1.1 Quartz

Quartz is the most suitable material for vessels in which acidic wet digestion of biological samples is to be carried out (i.e. also for the commonly used inserts for pressure digestion sample vessels).

Quartz is composed of at least 99.8% SiO_2. Therefore significant contamination of the sample material is not to be expected in most cases. Its surface remains smooth and non-porous, even after frequent use, and thus losses due to adsorption on the walls are largely eliminated.

Quartz exhibits a high resistance to all inorganic acids with the exception of hydrofluoric acid and concentrated phosphoric acid. Even mixtures of mineral acids with oxidising agents (including halogens) do not attack quartz. However, it is attacked by strong alkalis to form silicates. While volatile SiF_4 is generated by the reaction with hydrofluoric acid, silicophosphoric acid appears as a white corrosion on the surface due to the action of concentrated phosphoric acid. In both cases the surface becomes rough, which poses a risk of loss of certain elements.

Quartz is insensitive to temperature fluctuations. Its thermal conductivity is low. The maximum working temperature is about 1200 °C.

3.1.2 Borosilicate glass

Borosilicate glass vessels (e.g. Pyrex, Vycor, Duran) offer a feasible alternative for all wet digestion procedures using mineral acids for which quartz vessels are unsuitable, at least in the mg/L range and at higher concentrations. This does not apply to ultra-trace analysis in the µg/L range and at lower concentrations.

Borosilicate glass contains 81 to 96% SiO_2. The risk of contamination decreases as the SiO_2 content rises. The glass vessels have a smooth surface when new. All types of glass suffer strong attack from alkalis. Their resistance to acids depends on the type of acid as well as on its concentration. The following classification serves as a guideline for the resistance of this material:

$$HCl > H_2SO_4 > H_3PO_4 > HClO_4 \simeq HNO_3$$

Borosilicate glass has a low thermal conductivity and it is sensitive to extreme temperature fluctuations. The softening point of borosilicate glass is about 800 °C.

Before the first digestion, this type of glass must be conditioned with the acid to be used under the conditions for the digestion. If the glass becomes turbid after being used several times, it may no longer be used, as the rough surface poses the risk of losses due to adsorption.

Normal glassware is not suitable for wet digestion methods.

3.1.3 Polytetrafluoroethylene (PTFE)

PTFE is used as the vessel material for pressurised digestion systems, and in modern devices it often forms the casing for the quartz sample vessel (e.g. Seif apparatus). It is particularly suitable for digestions in which hydrofluoric acid has to be applied. The material has a smooth surface with micro-pores. The surface has paraffin-like properties, which means that it is practically wet-resistant to solutions. PTFE is almost completely resistant to acids and bases and also to mixtures of acids and bases with oxidising agents. However, as a result of the process by which it is manufactured – sintering of particles of the starting material under pressure – PTFE has pores in its structure. Loss of analytes can occur due to diffusion into the pores. In addition, these pores can cause undesired memory effects.

The maximum working temperature is about 200 °C. Above this temperature the material begins to flow and loses its mechanical strength. Therefore ashing devices made of PTFE must be encased in a jacket of heat-resistant, mechanically rigid material, as undesirable deformation can occur in the long term, even at temperatures below 200 °C. The vessel material usually poses no threat of contamination to the analytical sample.

3.1.4 Perfluoroalkoxy polymer (PFA)

This material, with properties very similar to those of PTFE, has a very smooth and sealed surface, therefore it does not have the disadvantages of PTFE, such as memory effects, etc. Its heat-resistance is comparable to that of PTFE. The material is highly recommendable when working with hydrofluoric acid [2].

3.1.5 Glassy carbon

Glassy carbon offers several advantages over other materials, such as high thermal resistance, resistance to acids, low memory effects. However, it is not easily processed mechanically, so that vessels are only commercially available in certain sizes. In addition, contamination is to be expected for some elements.

4 Digestion reagents

The reagents required for the digestion of biological material can be divided into three categories:
- oxidising agents,
- solubilisers and neutralisation agents,
- catalysts.

These reagents can contaminate the sample, depending on the content of the elements to be analysed. The concentrations of these elements given by the manufacturers of these reagents are maximum limits that in many cases depend on the detection limit of the analytical method used. Therefore different aspects must be taken into account when selecting and assessing reagents.

1. The ratio between the content of the elements to be determined in the analytical sample and the maximum possible contamination caused by the addition of the reagent should be as high as possible. If the possible contamination is in the same order of magnitude as the concentration of the element to be determined in the sample, then the reagent should not be used. Most reagents are available in different grades of purity. In many cases the use of reagents of the highest purity (e.g. Suprapur) is not necessary if the expected contamination of the analytical sample due to a reagent of lower quality (e.g. p.a. grade) does not cause interference to the analytical result. Reagent blank values must be checked for all the reagents used for ashing. This test must also be performed on different batches of the same reagent.
2. The reagents must also be assessed from the point of view of the secondary products that are formed from them. The volatility of the reagents and the possible interference that they may cause to the subsequent determination method must also be taken into account. Thus for example, HCl is less suitable than HNO_3 for ICP-MS on account of the possible polyatomic interference with other atoms caused

4 Digestion reagents

by Cl. In the case of AAS using the graphite furnace technique, the graphite tube has a considerably shorter useful life when HCl is used.

3. In addition to the trace elements to be determined in biological materials, there are always relatively large amounts of other elements (Na, K, Mg, Ca, etc.). If these elements form poorly soluble compounds with the digestion reagent, the analytes may be entrapped in the precipitate, making them inaccessible to determination.

4.1 Oxidising agents

4.1.1 Nitric acid

Properties: Boiling point 122 °C (69.2%)
Commercially available in a high grade of purity
Usable for all atomic and mass spectrometric techniques

Nitric acid is a good oxidising agent. In addition to its oxidising effect, it has good solvent properties for most of the biological material in question. The reaction products of ashing with HNO_3 are volatile, and the metal salts are, with few exceptions, readily soluble. On account of its relatively low boiling point, excess acid can be relatively easily and almost completely removed after completion of the ashing process. In comparison to the other ashing reagents, nitric acid shows the lowest values for contamination. However, this is highly dependent on the grade of HNO_3 used (pro analysi, Suprapur, Ultrapur, etc., subboiling distilled).

A disadvantage is the possible formation of stable nitro bodies, which can be completely decomposed only at higher temperatures, and then usually by burning off. This may lead to substance loss. The formation of explosive gaseous mixtures has never been observed. The reaction rate can be considerably accelerated by suitable additives. Relatively large amounts of nitric acid are required when working in open systems without reflux cooling, due to its high volatility.

4.1.2 Hydrogen peroxide

Properties: Boiling point 102 °C (30%)
More powerful oxidising properties as the milieu becomes more acidic
Commercially available in a high grade of purity

Hydrogen peroxide is a good oxidising agent for many organic substances. H_2O_2 is seldom used as an oxidising agent in its pure form. In most cases a mineral acid, normally HNO_3, serves as a solvent and proton donor. The reaction products are water and CO_2. Excess reagent is simply removed from the solution by boiling. Catalysts (e.g. Fe^{2+}) are frequently added when ashing is carried out with H_2O_2.

The content of potential contaminants is low, comparable with the levels present in nitric acid, but in this case too it depends on the grade used.

H_2O_2 can contain stabilisers (e.g. Ti compounds, H_3PO_4) that can have a detrimental effect in the digestion solution and on the analytical method. Digestion with H_2O_2 may lead to the formation of peroxo compounds of organic substances which can decompose in an explosive manner. Therefore appropriate safety precautions (closed fume cupboard, safety goggles) must be taken.

4.1.3 Potassium permanganate

Potassium permanganate is used for the decomposition of biological materials only in acidic aqueous solution. It is a strong oxidising agent. $KMnO_4$ is available in various grades of quality. Its content of contaminating trace elements is high. Its use is therefore restricted to wet digestion for the determination of mercury in biological materials. $KMnO_4$ may never be used as an oxidising agent in concentrated acids, as highly explosive Mn_2O_7 is generated.

4.1.4 Perchloric acid

Properties: Boiling point 203 °C (72.4%)
Strong oxidising agent
Explosive properties (!)

Concentrated perchloric acid at an elevated temperature is the strongest known oxidising agent for organic material. Due to its hazardous nature perchloric acid should never be used alone for the digestion of biological materials. However, it is normally essential as an additive to HNO_3 to increase the oxidation potential e.g. for the subsequent electrochemical determination of trace elements [3–5].

Moreover, with the exception of the Suprapur quality (or equivalent grades) perchloric acid contains considerable quantities of contaminating elements. This acid may never be used alone. It must be diluted with a suitable solvent, in general nitric acid. Excess acid may be removed by evaporation only when no more oxidisable substances are present. Working with $HClO_4$ poses the risk of explosion. Therefore the necessary precautions (explosion protection shield) must be taken.

4.2 Solubilisers and neutralising agents

4.2.1 Sulphuric acid

Properties: Boiling point 339 °C (98.3%)
Raises the boiling point in acid mixtures
Formation of insoluble sulphates

4 Digestion reagents

Affects the sample inlet system, e.g. in ICP-AES or ICP-MS, due to its high viscosity

Sulphuric acid has a very good dissolving capacity for many organic substances and has a strong dehydrating effect. For those reasons it is frequently used as an additive to oxidising acids. At high temperatures it also exhibits certain oxidative attributes. As a result of its relatively high boiling point it increases the oxidising potential of an acid mixture in closed systems. However, it is difficult to remove sulphuric acid from the digestion material, and interference can result. In many cases the residual acid must be neutralised using bases, which can represent an additional source of contamination. Many sulphates (e.g. $CaSO_4$) are poorly soluble and may lead to losses of the analyte due to occlusion.

H_2SO_4 is also available as a 96% solution in various grades of purity. The guaranteed content of trace elements in the Suprapur quality (or an equivalent) is less than that of other grades of purity by a factor of approximately 10 to 20.

Sulphuric acid is used in combination with hydrogen peroxide, e.g. for the digestion of small quantities of tissue samples (biopsies). This digestion can be carried out very rapidly within only a few minutes in an open vessel.

4.2.2 Hydrochloric acid

Properties: Boiling point 110 °C (20.4%)
Complexing agent for many precious metals
Formation of volatile chlorides (As, Sb, Pb, etc.)
Not advisable for use in graphite furnace AAS

HCl has no oxidising properties and serves only as a solubiliser or neutralising reagent. Most chlorides are readily soluble, and excess acid can be simply removed by boiling due to its low boiling point.

Its content of contaminating trace elements is higher than the equivalent grade of purity of nitric acid, but lower than that of perchloric acid. Its quality can be decisively improved by "subboiling distillation".

Hydrochloric acid is often used for the dilution of serum samples.

4.2.3 Ammonia

Ammonia is often employed to neutralise excess acid after wet digestion has been performed. Concentrated solutions of ammonia are available in Suprapur and p.a. grades. The content of contaminating trace elements is 20 to 100 times lower in the Suprapur grade. Poorly soluble compounds can only be formed with perchlorates. Excess ammonia can be simply removed by boiling. Gaseous neutralisation can also be achieved in modern appliances (e.g. GPN, Paar, Austria), for example by passing argon through an ammonia solution. This procedure has the advantage that no change in the volume of the sample occurs [6].

5 Digestion techniques

5.1 Combustion with oxygen in a closed system

5.1.1 Combustion in autoclaves

The homogenised investigation material (up to approx. 25 g) is dried by means of electrical heating in a quartz bowl in a closed pressurised vessel. Then the system is filled with oxygen (up to a pressure of 30 bar), and combustion is initiated using an incandescent wire. After being cooled, the pressurised vessel is opened and the resulting solution is drawn into a measurement vessel [7, 8].
(See Appendix: Example 1)

5.1.2 Schöniger combustion to determine iodine in biological material

The simple "Schöniger combustion" without the use of pressure offers an alternative that is particularly suitable for the determination of iodine in biological materials [9]. This procedure requires a normal Erlenmeyer flask with a ground-glass stopper, into which a Pt wire has been fused. To carry out the combustion the flask is filled with oxygen, and the previously ignited sample (in a filter paper) is introduced into the flask and burned. The special difficulty in determining iodine is that a normal acid digestion is not practicable, as most of the iodine present is converted to volatile iodine and not oxidised to the iodate, which leads to considerable losses, even in a closed system. The high oxidation potential required for rapid oxidation of iodide → iodine → iodate, which is achieved in the high-pressure digestion systems to be discussed later, is not available. Combustion of the sample in oxygen and the subsequent uptake in a basic medium (NaOH or tetramethylammonium hydroxide (TMAH)) prevents such losses. As a rule, TMAH is preferable due to its lower content of impurities and the lower stress caused to the ICP (strong cooling effect due to Na) and to the interface (lower salt burden).
(See Appendix: Example 2)

5.2 Wet digestion

Wet acidic digestion has proved especially successful for the determination of metals in blood, urine and tissue. These techniques are certainly the most widely used and are almost universally applicable. Samples may be digested in "open" or "closed" systems with different forms of "heat supply" (conventional or microwave heating).
Wet digestion of biological material involves the oxidative destruction of the organic matrix by means of suitable oxidising agents (such as HNO_3, H_2O_2, etc.) to form volatile reaction products, such as CO_2, H_2O, SO_2 and NO_x, while the ionic constituents are simultaneously dissolved.

5 Digestion techniques

The oxidation potential necessary to destroy the organic matrix is achieved by
- the oxidising acid,
- the temperature, and
- the pressure.

These three parameters determine the oxidation potential available for the digestion, and therefore the "quality" of the digestion solution. This is of primary importance for the "accuracy" of the analytical result.

5.2.1 Open systems

Wet digestion methods have proved advantageous over dry ashing as a result of the considerably lower reaction temperatures required for the process. This is largely due to the boiling points of the digestion reagents.
As the digestion is carried out in an acidic medium and in the liquid phase, losses due to adsorption or reaction with the vessel wall are minimised. However, losses due to volatilisation of elements must be taken into consideration. There is a risk of possible contamination due to the reagents used. Therefore reagent blank values of the systems must always be determined with the analytical series. On the one hand, the generally shorter ashing time has proved advantageous, but on the other, this technique is labour-intensive entailing a great deal of manual work and constant monitoring of the digestion.

5.2.1.1 Digestion reagent systems

The most frequently used digestion systems described below are as follows:
- HNO_3/H_2O_2 or Fenton's reagent
- HNO_3/H_2SO_4
- HNO_3
- $KMnO_4/H_2SO_4$
- $HNO_3/HClO_4$.

These individual systems and their mode of action are described below and examples of the relevant digestion methods are given (in the appendix). However, these should not be regarded as obligatory in the form described here for every analysis, they must be modified to suit the individual requirements for the type of material, the quantity to be digested as well as the expected trace element content and the type of analytical method to be used.
It is absolutely essential to wear safety goggles when working with concentrated acids or alkalis. In the case of chemical burns the affected part of the body must be rinsed immediately under running water.
In all wet digestion procedures large quantities of toxic gases and vapours such as NO_x, SO_2, SO_3, HNO_3, HCl, etc. are released. It is therefore essential to ensure that these toxic substances are removed without danger by using suitable digestion de-

vices or by carrying out the work in fume cupboards with efficient extraction systems. Explosive mixtures can be formed when certain reagents are used. In such cases suitable protective devices (wire meshing, acrylic glass panels) are essential.

Nitric acid/hydrogen peroxide or Fenton's reagent

This reagent system for the wet digestion of biological materials is particularly suitable for the determination of trace elements.

HNO_3 and H_2O_2 both act as oxidising agents in this system. While HNO_3 reacts as described in Section 4.1, the reaction of H_2O_2 proceeds largely via a radical mechanism (OH radicals). This reaction (Fenton) is catalysed by the iron that is always present in biological material. If necessary, Fe(II) solution must be added to urine so that the solution contains approximately 0.001 M of Fe ions. HNO_3 also acts as a solubiliser (hydrophilicity). After the digestion, many metals are present as readily soluble nitrates. Volatile metal compounds are not formed.

The use of this type of digestion is to be recommended, especially in open systems. No interference to subsequent analytical procedures has been observed. The only exception is the use of electrochemical methods, as nitric acid causes general interference in this case.

(See Appendix: Example 3)

Nitric acid/sulphuric acid

This reagent system for the wet digestion of biological materials is also particularly suitable for the determination of trace elements.

The combination of nitric acid and sulphuric acid has proved even more successful than nitric acid alone for the wet digestion of most biological materials. The actual oxidising agent is nitric acid. The addition of sulphuric acid has a dehydrating effect, promotes the formation of NO^+ and NO_2^+, and serves as a solvent, especially at higher temperatures. NO^+ and NO_2^+ are particularly strong oxidising agents. The addition of sulphuric acid enables the use of a higher temperature, whereby any nitro compounds formed are completely destroyed.

When sulphuric acid is used, there is a risk of insoluble sulphates being formed (e. g. Pb sulphate), and therefore H_2SO_4 is not a suitable digestion agent for every analyte. The advantage of using sulphuric acid in a closed system is that the temperature increases and thus a higher oxidation potential is achieved.

(See Appendix: Example 4)

Nitric acid

The use of nitric acid alone for the digestion of biological materials in an open system is restricted to special applications.

5 Digestion techniques

Nitric acid acts as an oxidising agent in the presence of most organic substances, whereby partial to complete digestion yielding volatile reaction products is generally achieved. Concentrated HNO_3 contains low concentrations of the nitrosyl cation NO^+, which enables electrophilic substitution in organic material. This increases the hydrophilicity.

After wet digestion, most metals are present as readily soluble nitrates. Volatile metal compounds are not formed. The oxidising and solvent effect depends on the acid concentration and the digestion temperature.

If the acid is evaporated too early before the matrix is completely digested, brown to black residues are left. Further addition of acid does not achieve complete digestion of these residues. Crusts of stable nitro bodies can be formed on the glass wall during the digestion. Further addition of acid tends to cause spontaneous combustion of these bodies, whereby element losses are possible. Swirling round occasionally prevents the formation of these crusts. No explosive mixtures are generated.

Substances containing fat can only be digested with difficulty using HNO_3.

The use of HNO_3 alone for the wet digestion of biological materials requires longer digestion times than those needed for acid mixtures. In addition, the risk of contamination is increased by the larger quantities of acid required. Due to these factors it is advisable to carry out digestion with nitric acid alone under elevated pressure, as these drawbacks are thus avoided.

Potassium permanganate/nitric acid/sulphuric acid

The treatment of biological materials with potassium permanganate in acidic aqueous solution serves exclusively as a preparation for the subsequent determination of mercury by means of flameless atomic absorption spectrometry.

After the reaction, all the mercury (also mercury from organic compounds) is present in ionic form (Hg^{2+}). In contrast, the organic part of the matrix is only incompletely oxidised. If a mixture of nitric acid and sulphuric acid is used and the reactants are heated to 60 to 80 °C, the organic constituents (with the exception of fat) are largely decomposed to compounds that are readily soluble in water. This prevents volatile compounds that exhibit absorption at 254 nm from becoming gaseous in the subsequent determination, which would lead to a higher background value.

As already mentioned, this procedure is used exclusively for Hg determination by means of flameless AAS. As this digestion is carried out in an open system, its importance is steadily declining compared with closed systems under pressure and at high temperature. Here the addition of potassium permanganate is only advisable after the digestion in order to stabilise the solution.

(See Appendix: Example 5)

5.2.2 Closed systems

5.2.2.1 Digestion systems under medium and high pressure

Digestion of biological matrices at pressures currently up to a maximum of about 120 bar is carried out in vessels made of PTFE, PFA, glassy carbon or quartz that are inserted into pressure vessels made of stainless steel or other metals [10–16]. The sample together with the digestion reagents are subjected to a temperature programme in the pressurised digestion device. Initially, the more reactive sample constituents are decomposed at lower temperatures, then finally the sample is completely digested at temperatures above the boiling point of the digestion reagents.

In the meantime reagent systems and temperature programmes for trace analysis have been optimised for various biological materials [11, 14, 16].

The most commonly used vessels for pressurised digestion are made of PTFE with a pressure jacket of metal, which permits temperatures of up to 150 °C [17, 19]. Above this temperature PTFE begins to soften. Therefore the use of glassy carbon and quartz vessels, which are thermally very stable and permit digestion with most reagents at temperatures above 300 °C, is increasingly common [18].

The most widely used pressurised digestion devices are suitable for relatively low levels of overpressure (approx. 6 to 12 bar), PTFE or PFA sample vessels (possibly with quartz inserts) (e.g. "Seif apparatus"), and low temperatures (at most 180 °C) [17]. This digestion technique does not always achieve complete digestion of the biological material, especially if nitric acid is used alone (incomplete digestion causes interference e.g. with electrochemical detection methods [20, 21]). However, the quality of the digestion solution is certainly sufficient for atomic spectrometric or mass spectrometric analytical methods. The typical residual carbon values are in the range of up to 10%, and this is not sufficiently effective for some analytical steps, such as matrix separation, concentration or hydride formation, or for electrochemical procedures.

For this reason the oxidation potential has to be enhanced, i.e. an increase in temperature and pressure is necessary, which also entails the use of more resistant materials such as quartz or glassy carbon.

The samples or their reaction products that cause most problems are [22]:
- Poly-unsaturated fatty acids that can only be converted to the *cis*-1,2-cyclopropanedi32carboxylic acid with HNO_3 at 180 °C. This ring is not cleaved at 180 °C.
- Certain amino acids:
  a) Phenylalanine, which yields various nitrobenzoic acids as stable end products.
  b) Histidine, which has an aromatic carboxylic acid as its stable end product.
  c) Tryptophan, which forms several stable aromatic compounds, some of which are carboxylic acids.
  d) Methionine, which is converted only as far as the stable methanesulphonic acid.

After a pressurised digestion at only 180 °C and at low to medium overpressures, these stable reaction products cannot be evaporated as volatile compounds with nitric acid (at about 130 °C). They remain in the digestion residue and cause interference e.g. in the final voltammetric analysis.

5 Digestion techniques

Thus stable reaction products, such as those of tryptophan, are regarded as maximum dampers that also increase the flow of hydrogen. The nitrobenzoic acids formed from phenylalanine generate strong interfering signals that impede the determination of Cd, Pb and Cu in particular. They also increase the flow of hydrogen and thus impair the signal/background ratio, e.g. for the analysis of Zn.

A practically complete mineralisation of the above-mentioned reaction products is only possible at approx. 300 °C with HNO_3. Element determination without interference is only possible after evaporation of the dissolved nitrogen oxides, which are known to cause high background noise in electrochemical determination methods. However, these nitrogen oxides that interfere with voltammetric measurements are decisive for the complete decomposition of the organic compounds. Therefore a sufficiently high concentration of NO_2 must be present in the digestion system. Addition or an excess of NO_2 (e.g. generated by fuming nitric acid) enhances the decomposition rate. Such a system must therefore be airtight to maintain the pressure of NO_2, but this is only achieved by the high-pressure digestion systems.

The characteristics of a high-performance digestion procedure are therefore:
- Low consumption of reagents;
- The ratio between the vessel surface and the sample quantity should be as small as possible;
- Pure and chemically inert vessel material;
- The use of closed systems;
- Complete breakdown of the organic matrix.

Pressurised digestion devices are increasingly being equipped with microwave heating (e.g. Pressurized Microwave Decomposition (PMD), Multiwave, etc.) as an alternative to conventional heating (e.g. High Pressure Asher (HPA)).

These decomposition systems permit pressures up to approx. 130 bar (HPA) and temperatures up to about 320 °C, which considerably enhance the oxidation potential. The "quality" of these digestions, i.e. destruction of the organic matrix as completely as possible with typical residual carbon values of <1%, even meets the requirements of electrochemical methods and is adequate for solving special problems such as determination of total arsenic in samples of marine origin or in urine, where in some cases very stable organic As ligands are present [21].

The latest developments, specially for the decomposition of body fluids, are flow-through systems with microwave heating at pressures up to 35 bar, and UV digestion systems [23–25].

With flowthrough systems it is possible to use an autosampler that has an on-line connection to a suitable measurement device such as flame AAS or ICP-AES. The digestion times for body fluids are in the range of only one to several minutes.

The UV digestion systems, in which the organic material is destroyed by UV radiation, are especially suitable for the digestion of dissolved organic materials in aqueous sample solutions (digestion is also sufficiently effective for electrochemical determination methods). With the addition of H_2O_2 or HNO_3 these systems can also be used for digesting body fluids. In this case the digestion solution is of sufficiently high quality for atomic spectrometric determination methods, but not for electrochemical procedures.

Apparatus for pressure digestion must be secured against operational errors and the build-up of explosive pressures, i.e. when a pressure limit value (depending on the design of the metal jacket) is reached (50 to 200 bar), it must be possible to release the gases from the apparatus without danger. Depending on the manufacturer, this is achieved either by spiral or plate springs pressing on the cover of the digestion vessel (principle: Tölg [11], manufacturers: Berghof, Perkin Elmer), by fitting rupture discs (manufacturers: Paar, Uniseal, Berghof) or by a suitable spring design outside the vessel connected with conical Teflon stoppers (Seif system, Seif-Aufschlusstechnik) [17]. The heating of the pressure reactors can be carried out in a metal block (aluminium or brass) and also in air thermostats.

When digestion is carried out under pressure, the safety precautions specified by the manufacturer for working with autoclaves must be observed.

Systematic studies on the pressure-temperature behaviour in systems of the above-mentioned designs [26] showed that, depending on the matrix, pressure peaks (pressure waves) of considerable magnitude can occur which are the cause of rupture or gas escape. Thus a pressure peak of 15 bar was measured during the ashing of 1 mL whole blood with HNO_3, and the ashing of 1 g of dried fish filet caused a peak of 90 bar. This demonstrates the necessity of carrying out careful preliminary experiments with each matrix for which no reagent system or temperature programme has yet been established.

In addition to the absorption and adsorption effects on vessel walls, which are always to be observed in the trace analytical range, and the risk of contamination due to insufficient cleaning of vessels and impurities in the chemicals, it must be taken into account that the vessel surface that comes into contact with the digestion reagents undergoes an ageing process. Electron microscopic investigations of PTFE and glassy carbon surfaces confirm these surface changes [15]. Therefore timely replacement of digestion vessels is advisable in order to ensure exact results for trace analyses in the ng/g range. The use of quartz vessels (e.g. HPA, PMD or Seif apparatus) is recommended in this range. Before using new vessels, they should be checked for blank values under the relevant digestion conditions.

Polarographic measurements in digestion solutions and gas chromatographic analyses of the gaseous mixture above the digestion solution to determine the carbon ratio show, as already mentioned, that pressure digestion with HNO_3 (at low to medium overpressures) and at a maximum of 170 °C does not always achieve complete mineralisation of the organic material [20, 21, 26]. This must be taken into consideration when devising analytical procedures. Thus subsequent oxidation with $HClO_4$, H_2O_2 or mixtures of $HClO_4$ and H_2O_2 with H_2SO_4 or the use of high-pressure systems is frequently unavoidable, especially for subsequent measurement with electrochemical methods or for determination of total arsenic [20, 21].

Digestion under pressure is primarily suitable as a preparation for determinations by means of atomic absorption and emission spectrometry, and also for electrochemical methods when high pressures and certain acid mixtures are used. However, it can also serve as preparation for other determination methods. Pressure digestion methods are suitable for determination of trace elements in biological material. It is particularly recommendable if only small sample amounts are available.

In principle, pressure digestion has the following advantages over open systems:
- The use of digestion temperatures above the boiling point of the digestion reagents shortens the digestion time.
- The reagent consumption is lower than in open systems without reflux. Thus reagent blank values are greatly reduced.
- Contamination or losses of the analytical material on the vessel surface are minimised as a result of the small dimensions of the ashing vessel.

The cost of the apparatus and the digestion vessels can be regarded as a disadvantage of this method. If not handled correctly, dried materials may react so vigorously that losses can occur due to blow-out of the safety system. However, this can be easily ascertained from the quantity of acid that remains. If the digestion conditions are correctly selected, excellent recovery rates are achieved as radiochemical experiments have shown [15, 16, 26].

If no such pressure digestion systems are available, the digestion procedure under low overpressure briefly described in the following section offers a satisfactory alternative, at least for a subsequent atomic spectrometric or mass spectrometric determination.

5.2.2.2 Digestion under low overpressure

Blood in quantities up to 0.3 mL can be converted within one to two minutes into a relatively clear solution in which metals such as Pb, Cd, Ni, etc. are present as ions. This is achieved with a mixture of dilute nitric acid and hydrogen peroxide while heating to 115 °C in test-tubes with screw caps (permissible inner pressure 5 bar) [27]. The primary oxidising agents are the OH radicals that are formed in the presence of iron(II) ions.

Potassium permanganate in a solution of sulphuric acid/nitric acid is used as the digestion agent in order to break down urine or blood for the determination of mercury.
(See Appendix: Example 6)

6 Current "state-of-the-art" for digestion procedures

The current "state-of-the-art" in digestion technology for organic materials can be summarised as:
1. Wet ashing in closed systems
    1.1. Digestion in PTFE or PFA sample vessels (possibly with quartz inserts) at medium overpressures (approx. 15 bar) with conventional or microwave heating.
    (See Appendix: Example 7)
    1.2. Digestion in closed PTFE, PFA, quartz or glassy carbon vessels with microwave heating at high pressures (approx. 85 bar)
    e.g. PMD (from Paar, Austria)

Multiwave (from Paar, Austria)
Microwave Acid Digestion Bomb (from Parr, USA)
(See Appendix: Examples 8 and 9)
  1.3. Digestion in closed quartz vessels at high pressures with conventional heating
    e.g. HPA (from Paar, Austria)
    (See Appendix: Example 10)
2. Flowthrough digestion system with microwave heating and pressures of approx. 35 bar (especially suitable for the digestion of body fluids)
    e.g. Auto-Flow (from Berghof)
    (See Appendix: Example 11)
3. UV digestion system
    e.g. UV 1000 (from Kürner Analysentechnik, Germany)
    (See Appendix: Example 12)

The flowthrough digestion systems are particularly suitable for automation of digestion and for on-line coupling with measurement instruments, such as flame AAS, ICP-AES, etc. [23].

The UV digestion systems are especially effective for decomposing dissolved organic materials in aqueous sample solutions (such as water, waste water, fruit juice, etc.). The material is destroyed by UV radiation. It is necessary to add H_2O_2 or also HNO_3 to achieve satisfactory results for the digestion of more complex matrices, such as urine, serum, etc. [24, 25].

7 Appendix

7.1 Examples of procedures

7.1.1 Example 1: Digestion under oxygen in a closed system

Equipment
Pressure digestion device for the rapid digestion of biomaterials in the presence of oxygen (e.g. "Bioklav®", from Siemens AG, Karlsruhe, Germany)

Chemicals
Oxygen, subsequently purified
Nitric acid, Suprapur (or equivalent) or subboiling distilled (diluted 1:2)
Ash-free round filter (e.g. from Schleicher + Schüll)

Sample preparation
10–25 g of the biological material to be investigated are weighed in a quartz vessel. Tissue samples can be homogenised before weighing.

Procedure
The sample to be digested, which must have a water content of at least 20%, is placed in the sample vessel between two round filters. (The residual water content facilitates the weighing process and the spatial stabilisation of the sample between the filter papers – but it is not absolutely necessary!) After a strip of filter paper has been attached to the Pt ignition wire for ignition and the sample dish has been inserted into the pressure vessel, the autoclave and the valves are closed. The analytical material is then pre-dried for 10–30 min. On completion of drying, oxygen is supplied to the autoclave up to a pressure of 35 bar, and the material is ignited. When the reaction is completed, the overpressure is released slowly and the pressure vessel is opened, the cover is rinsed with a small amount of H_2O, the sample dish is taken out, the heating device and the outside of the sample dish are rinsed with H_2O and the condensate is removed from the bottom of the device.
Then 4 mL of the HNO_3 solution are added to the contents of the sample dish, it is heated and the dish is rinsed again using 10 mL H_2O. After rinsing again, the rinsing solutions are combined. The total volume is 100 mL at most.

7.1.2 Example 2: Schöniger combustion for the determination of iodine

Equipment
Erlenmeyer flask, 500 mL
Ground-glass stopper with a Pt wire fused into it

Chemicals
TMAH solution (2.5%). The product from TAMA Chemicals, Japan (commercially available as a 25% solution) is to be recommended due to its extremely high purity.

This method is especially suitable for ashing solid samples (also body fluids after e.g. freeze-drying) [9].

Procedure
20–30 mg of the dried sample are weighed on an ash-free filter paper (e.g. Whatman 542) and then wrapped up in it. The wrapped sample is fixed to the Pt wire.
5 mL TMAH solution (2.5%) are pipetted into the Erlenmeyer flask, and the flask is filled with oxygen. The filter paper is ignited and immediately introduced into the flask. The ground-glass stopper is pressed into the opening of the flask. The sample burns in oxygen without a residue.
The closed flask is left to stand for approx. 1 h and the condensed vapour is rinsed from the walls 2 to 3 times during this time by swirling around the TMAH solution previously placed in the flask.
The flask is subsequently opened and rinsed together with the stopper and the Pt wire with a further 5 mL of the TMAH solution.
This solution is very suitable for the determination of iodine by means of ICP-MS. Other trace elements can also be determined (also using another solvent) if for example only small amounts of sample are available for analysis.

7.1.3 Example 3: Open ashing with HNO_3/H_2O_2 (Fenton's reagent)

Equipment
Glass beaker or quartz flask
Heating bath or heating plate

Chemicals
Nitric acid, Suprapur (or equivalent) or subboiling distilled
Hydrogen peroxide, p.a. 30%
Iron(II) ammonium sulphate, p.a. (0.1 M solution)
Hydroxylammonium chloride, p.a. (10% solution)

Sample preparation
Sample preparation is not necessary for wet ashing of blood samples. However, it is advisable to evaporate urine to approx. 1/5 of its starting volume. Homogenised tissue can be digested as a moist substance.

Procedure
The addition of HNO_3 is the first step in the digestion process. If necessary, an appropriate amount of Fe(II) solution is added. The first oxidation phase, accompanied by the generation of gases, is allowed to proceed, whereby the contents of the vessel should be carefully swirled around occasionally. Then the digestion solution is heated until it is golden yellow. The reaction mixture is allowed to cool, H_2O_2 is added slowly and the mixture is evaporated with occasional swirling. When the concentrate begins to turn a dark colour, more H_2O_2 is added and evaporation is repeated. If necessary, this step must be repeated until the concentrate remains colourless so that a moist, colourless crystalline residue is left when evaporation is contin-

7 Appendix

ued. In digestion processes for the determination of arsenic the excess H_2O_2 must be decomposed by boiling with hydroxylammonium chloride solution.
The following examples can serve as guidelines for the amount of sample material and reagents required:

Urine: 10 mL HNO_3 and 2 drops of Fe(II) solution are added to 10 mL of the evaporated urine sample. The procedure is carried out as described above until completion of the first digestion phase. Then 5 mL H_2O_2 are added as described above. If necessary, the addition must be repeated until complete digestion has been achieved.

Blood: 5 to 10 mL HNO_3 are added to 2 mL blood. The procedure is carried out as described above until completion of the first digestion phase. Then 5 mL H_2O_2 are added as described above, and if necessary this is repeated until digestion has been completed.

Tissue: 10 mL HNO_3 are added to 10 g of coarsely comminuted tissue or 1 g of dried substance that has been previously moistened with H_2O. Then H_2O_2 is added in 5 mL portions. The procedure is the same as that described for urine and blood.
Providing the analytical requirements are met, smaller sample amounts can be processed, naturally also with correspondingly smaller quantities of reagents. Thus, for example, biopsy samples, which are generally available in amounts of <10 mg, are digested with 200 µL acid and approx. 300 to 500 µL H_2O_2, whereby a period of only a few minutes is required.

Safety precautions
If the material is not completely digested, peroxo compounds present in the solution can decompose spontaneously during evaporation. Therefore evaporation to dryness should be avoided.

Sources of error
The formation of carbon in the evaporated solution, recognisable by the dark brown coloration, must be avoided. Once elemental carbon has been formed, it cannot be further oxidised under the test conditions. Such samples must be discarded. As this reagent system tends to cause violent reactions as a result of the radical mechanism, the heating bath temperature should not exceed 130 °C.
As reagent blank value measurements have shown, contamination due to trace elements being carried over is very low.
The moist crystalline residue may still contain excess H_2O_2, which could cause interference to the subsequent analysis. After dissolution in H_2O, it can be removed by boiling (autocatalytic decomposition) or by addition of hydroxylammonium chloride solution.

Discussion
About 2 to 3 hours are required for this method, which is not very time-consuming, and it is sufficiently practicable for routine investigation, provided the reaction conditions are under good control.

This wet digestion method using H_2O_2 with minimum addition of acid of the "Suprapur" purity grade is superior to those methods requiring larger amounts of acid with regard to possible cross-contamination with trace elements.

7.1.4 Example 4: Open digestion with HNO_3/H_2SO_4

Equipment
Glass beaker or quartz flask
Heating bath or heating plate

Chemicals
Nitric acid, Suprapur (or equivalent) or subboiling distilled
Sulphuric acid, Suprapur (or equivalent)
Ammonia, Suprapur (or equivalent)

Sample preparation
Sample preparation is not necessary for wet ashing of blood. However, it is advisable to evaporate urine to approx. 1/5 of its starting volume. Homogenised tissue can be dried or processed in a moist state. The use of dry substance reduces the actual digestion time and the consumption of reagents.

Procedure
The initially added volume of HNO_3 depends on the amount of substance to be digested and also on the type of digestion vessel or apparatus. H_2SO_4 can be added in various ratios to the amount of HNO_3.
It has proved advantageous to use a ratio of approx. 10 parts of HNO_3 to one part of H_2SO_4. In general, 10 mL HNO_3 are added to 5 to 10 mL blood or concentrated urine. Then 1 mL H_2SO_4 is added slowly. The spontaneous reaction that starts after a short time is allowed to subside without supplying heat. If vigorous foaming occurs, 1 or 2 drops of an alcohol (e.g. ethanol) should be added. Then the digestion vessel is heated until there is a slight reflux of HNO_3 and kept at this temperature (120 to 150 °C) until the solution appears colourless or slightly yellowish and no more nitrogenous gases are generated. Then the temperature is increased to 250 °C until white fumes of SO_3 appear. This fuming-off may not be continued, as losses of the analyte elements can result. A colourless to slightly yellowish solution indicates the end of the digestion process. However, if the solution is dark, it must be cooled and then 1 to 2 mL HNO_3 must be added, the solution must be heated and evaporated until white fumes appear. This process must be repeated until a pale solution is obtained. If the residual sulphuric acid in the solution interferes with the subsequent determination procedure, it can be diluted with H_2O and neutralised with an excess of ammonia. Ammonia is easily removed by evaporation. The residue is soluble in H_2O.
These instructions must be varied for As determination. 20 mL HNO_3 are added to 5 to 10 mL blood or concentrated urine in a Kjeldahl flask, and 5 mL H_2SO_4 are slowly added. After the initial spontaneous reaction subsides, the mixture is heated. The further procedure is as described in the above instructions.

7 Appendix

Sources of error

After the digestion, the metals are present as sulphates or oxides. If the sulphates of the elements to be determined are poorly soluble or if large amounts of concomitant elements are present in the form of poorly soluble sulphates, this can considerably influence the subsequent determination.

Discussion

Wet digestion with HNO_3/H_2SO_4 is suitable for blood, urine and tissue. Most non-volatile trace elements can be determined in the digestion solution, even arsenic with slight modifications to the digestion instructions.

As the sample is never evaporated to dryness in the presence of H_2SO_4, the loss of elements due to nitro compounds burning off is not to be expected. The higher temperature of approx. 250°C serves to decompose the nitro compounds which would cause interference to the subsequent determination. With good regulation of the temperature of the heating bath, the digestion requires constant monitoring for 2 to 4 hours, depending on the type of sample and the apparatus used. Compared with the wet digestion using HNO_3 alone, less HNO_3 is required for the same quantity of biological material. As shown by reagent blank values, contamination with trace elements is low, but not negligible when p.a. or "Suprapur" grades of acids and ammonia are used. The higher temperature compared to digestion with HNO_3/H_2O_2 results in more effective digestion. However, the formation of insoluble lead sulphate during the determination of lead must be taken into account.

7.1.5 Example 5: Open digestion with potassium permanganate for the determination of Hg

Equipment
Erlenmeyer flask

Chemicals
Nitric acid, Suprapur (or equivalent) or subboiling distilled (5 M)
Sulphuric acid, Suprapur (or equivalent) (4.5 M)
Potassium permanganate, Suprapur (5% solution of $KMnO_4$ in H_2O)
Hydroxylammonium chloride (saturated, approx. 60% in H_2O)

Sample preparation
No special sample preparation is required for urine or blood. It is advisable to comminute tissue samples.

Procedure
10 mL urine or 2 mL blood or 1 g tissue, 5 mL 5 M HNO_3 and 5 mL 4.5 M H_2SO_4 are heated to 60–80°C in an Erlenmeyer flask covered with a watch glass for 15–20 minutes. After cooling, 10 mL 5% $KMnO_4$ solution are added and the contents are heated again to 60–80°C for 10 minutes. The violet coloration should also remain after heating, otherwise $KMnO_4$ solution must be added again. After the contents are

cooled, saturated hydroxylammonium chloride solution is added dropwise to the flask while it is being shaken vigorously until the solution just becomes colourless.

Safety precautions
This digestion does not require any special safety precautions. $KMnO_4$ solution may never be evaporated to concentrate it.

Sources of error
Although fats are not digested or only incompletely digested by this method, experience has shown that no loss of mercury occurs. If too little $KMnO_4$ solution is added (the digestion solution is already decoloured after heating) or if heating at lower temperatures is carried out for too short a time, readily volatile organic compounds can be formed which simulate a falsely elevated content of mercury due to their absorption at 254 nm.
It is important to comply with the upper temperature limits in order to avoid potential losses of mercury.

Discussion
The digestion of blood and urine with potassium permanganate for atomic absorption spectrometric determination of mercury is relatively simple and is practicable in large series.

7.1.6 Example 6: Digestion under low overpressure (<5 bar)

Equipment
Electrically heated test-tube block, steplessly adjustable
Screw-capped test-tubes (e.g. Sovirel No. 15 or No. 20), inner diameter 1.6 or 2.0 cm, length 10 to 20 cm, permissible inner pressure 5 bar.
Enzyme pipettes
Microlitre pipettes

Chemicals
Hydrogen peroxide p.a. 30% (Perhydrol)
Nitric acid, semi-concentrated, approx. 33%
Nitric acid 1 M
Sulphuric acid semi-concentrated, approx. 48%
Potassium permanganate solution 5%, in sulphuric acid, pH 1, prepared with $KMnO_4$ of a guaranteed maximum Hg content
Hydroxylammonium chloride solution 6%
n-Heptane for spectroscopy

Procedure
Digestion of blood with nitric acid/hydrogen peroxide
0.3 mL nitric acid (1 M) is added to 0.05 to 0.1 mL blood in a 10 mL test-tube with a screw cap, and the contents are thoroughly mixed so that no agglomeration occurs in the brown suspension that is created. Then 0.5 mL H_2O_2 is added and thoroughly

mixed. Finally, the sample is filled with distilled water, if necessary adding a metal standard, to a final volume of 1 mL. The test-tube is screwed shut, shaken and heated in the heating block to 110 to 115 °C. After 20 to 30 seconds the spontaneous decomposition reaction begins. This is recognisable by the formation of foam and the decoloration of the solution. The sample is left in the heating block for three minutes, during which it is shaken several times, then after cooling it is carefully opened and shaken until no more gas escapes from the solution. The digestion solution thus obtained is almost clear up to a volume of 100 µL blood, and can be used in this form for metal determination according to the standard addition procedure in graphite furnaces [28].

Digestion of urine and blood for the determination of mercury by means of flameless atomic absorption spectrometry

Urine: 0.5 mL semi-concentrated sulphuric acid, 0.5 mL semi-concentrated nitric acid, and 2 mL sulphuric potassium permanganate solution are filled into a screw-capped test-tube with a capacity of 20 mL. Then 1 mL of urine is added to this reaction mixture, the tube is screwed shut, the contents are well mixed and the solution is heated for approx. 20 minutes in the heating block. After cooling, 2 mL distilled water and 1 mL hydroxylammonium chloride solution are added, the latter in order to reduce the excess oxidising agent. This solution can be used directly for the determination of mercury by means of flameless atomic absorption spectrometry [29].

Even larger quantities of urine can be digested in the manner described here. A test-tube with a capacity of 40 mL is used for the digestion of 5 mL urine. In this case the reaction mixture consists of 5 mL semi-concentrated sulphuric acid, 5 mL semi-concentrated nitric acid, and 10 mL potassium permanganate solution. Reduction of the excess oxidising agent is achieved by adding 5 mL hydroxylammonium chloride solution.

Blood: Digestion of blood is carried out as described above. A test-tube with a capacity of 20 mL is used.

The following reagent solutions are used for 0.5 mL blood:
1 mL sulphuric acid, semi-concentrated, 48%
1 mL nitric acid, semi-concentrated, 33%
4 mL potassium permanganate, sulphuric solution.

After cooling, 0.5 mL distilled water is added, the test-tube is shaken, and then the excess oxidising agent is reduced using 1 mL 6% hydroxyammonium chloride solution.

Safety precautions
Digestion of urine and blood with potassium permanganate in acidic solution is not problematic under the given conditions. The pressures occurring during the reaction are far below the permissible limit of 5 bar.

Distinctly higher inner pressures occur during the digestion of blood with nitric acid/hydrogen peroxide. In certain cases, especially with older, frequently used closures, the screw caps can rupture. Therefore it is advisable to replace the closure after about 50 digestions. When highly concentrated nitric acid is used, the inner pressure increases greatly, so the reaction mixture should be prepared with a maximum concentration of 2 M nitric acid, as the use of concentrated nitric acid can lead to rupture of the vessels.

Sources of error
Interference has not been observed in the digestions described here. Impurities due to the reagents used are taken into account by analysing appropriate blank solutions. Used test-tubes must be cleaned with great care: heating the test-tubes filled with a 5% detergent solution for 30 minutes in the heating block at 100 °C and shaking several times has proved successful. Then they are rinsed with water and approx. 1.5 M nitric acid.

Discussion
The method described here is sufficiently effective for the subsequent determination of metal ions by means of atomic absorption spectrometry.
Screw-capped test-tubes are commercially available in many sizes and are definitely the least expensive digestion vessels under low overpressure. Their use is restricted by the inner pressure limit of 5 bar, but this is often sufficient to digest the sample amounts required for this determination method in a short time without losses. The described digestion of blood using nitric acid/hydrogen peroxide has been successfully used for the determination of lead in accordance with the standard addition procedure using a graphite furnace.
Digestion of urine and blood using acidic potassium permanganate solution in screw-capped test-tubes yields good results for the determination of total mercury.
This is exclusively a description of alternative digestion methods. In principle, pressure digestion with HNO_3 would be sufficient for all biological materials. Open digestion methods have the disadvantage of the risk of contamination due to laboratory dust and also through possible loss of readily volatile elements or compounds.

7.1.7 Example 7: Digestion under medium overpressures (<20 bar) using a "Seif apparatus"

The "Seif apparatus" is a simple pressure digestion apparatus operated at pressures of about 15 bar and a maximum temperature of about 180 °C. It consists of a metal block with 6 drilled holes into which the Teflon vessels with conical Teflon covers are inserted. These covers are pressed onto the vessels by springs to seal them. The pressure of the springs determines the maximum achievable pressure. Quartz glass vessels with quartz covers are inserted into the Teflon vessels. The samples and the acid are introduced into these vessels so that they only come into contact with the quartz. This prevents both contamination with impurities from the Teflon and the well-known "memory effects" shown by Teflon. The entire apparatus is placed in a drying cupboard at 180 °C to achieve digestion.
This procedure is suitable for all organic matrices. Up to 12 samples per day can be digested using this apparatus, as the time required per digestion run is about 6 hours.

7 Appendix

This digestion procedure is particularly suitable for the subsequent determination of trace elements by means of atomic or mass spectrometric methods. Care must be taken with regard to the "digestion quality" when electrochemical determination methods are to be used or when total arsenic is to be determined, e.g. in urine, as the oxidation potential may not be sufficient to destroy the entire organic matrix in this case.

Equipment

Pressure digestion apparatus according to "Seif" (Seif-Aufschlusstechnik, Unterschleissheim, Germany), available in 3 different sizes (6-sample block for maximum weighed-in amounts of 120 mg dry substance, 2-sample block for 500 mg dry substance, 1-sample block for 1000 mg dry substance) [17]. This apparatus can be operated up to a pressure of 12 to 15 bar.

The example given here is based on a 6-sample block (6 samples simultaneously). When the larger devices are used, the weighed-in amount of the sample must be adapted accordingly.

Drying cupboard

Chemicals

Nitric acid, Suprapur (or equivalent) or subboiling distilled

Procedure

1 mL or blood, serum, urine, milk or 100 mg of dried substance (correspondingly more of fresh substance) are weighed into the quartz vessels and 1 mL HNO_3 is added.

The quartz glass vessels are closed with the appropriate quartz covers, placed into the apparatus and then sealed.

The entire digestion block is treated in a drying cupboard at 170 to 180°C for approx. 6 hours.

After cooling, the clear sample solution is filled to the desired final volume with H_2O (ultrapure).

Interference and discussion

This procedure is suitable for all biological materials. The achievable oxidation potential is sufficient with regard to the residual carbon values (approx. 5–10% depending on the matrix) for all atomic and mass spectrometric analytical methods. Problems, which necessitate further treatment of the sample solution with $HClO_4$, arise only for electrochemical methods, such as voltammetry, as the residual carbon content should be <1% in this case.

7.1.8 Example 8: High-pressure digestion with microwave heating (e.g. pressurised microwave decomposition (PMD))

Microwave heating is used to heat the sample in this method. The sample vessels, also made of quartz, permit the use of pressures up to about 85 bar, the temperature reaches approx. 220 °C. These two parameters achieve a considerably enhanced oxidation potential. The digestion time is approx. 60 minutes (including the cooling phase). Two samples can be simultaneously inserted in the apparatus. Thus as many as 20 samples/day can be digested.
This digestion is universally suitable for all determination methods (after matrix-dependent modifications) [19].

Equipment
High-pressure microwave digestion apparatus "PMD" (Perkin-Elmer, Germany, A. Paar KG, Austria)
This apparatus can be operated up to a pressure of 85 bar and is suitable for simultaneous digestion of 2 samples.

Chemicals
Nitric acid, Suprapur (or equivalent) or subboiling distilled
Perchloric acid, Suprapur
Sulphuric acid, Suprapur

Digestion of blood (atomic and mass spectrometric detection)
1 mL blood and 2 mL HNO_3 are pipetted into a 35 mL quartz glass vessel (accessory), which is sealed in accordance with the manufacturer's instructions and inserted into the pressure vessel.
Digestion program: Power: level 6
                     Duration: 10 minutes
                     Cooling: 10 minutes
Then the clear solution is filled up to the desired final volume.

Digestion of urine (voltammetric detection, total arsenic)
1 mL urine, 1 mL HNO_3 and 0.5 mL $HClO_4$ are pipetted into a 35 mL quartz vessel.
Digestion program: Power: level 6
                     Duration: 20 minutes
                     Cooling: 20 minutes
The digestion solution is evaporated to dryness in a quartz crucible (excess HNO_3 or nitrogenous oxides must be eliminated from the solution) for a voltammetric determination. The residue is taken up in 20 µL $HClO_4$ and 10 mL H_2O.
Addition of approx. 250 µL sulphuric acid may be necessary to increase the temperature for blood or tissue samples.

Interference and discussion
No interference to subsequent analytical procedures has been observed. This procedure is suitable for all biological materials. Depending on the matrix and the analyti-

7 Appendix

cal method, acid mixtures – as the examples show – must be used to improve the effectiveness of the digestion.

7.1.9 Example 9: High-pressure digestion in the "Multiwave"

This digestion procedure (also with microwave-supported heating) is very similar to the PMD procedure, also with regard to the sample vessel, but in this case up to 6 samples can be digested simultaneously. A computer-controlled temperature program also permits optimum adaptation of the digestion parameters to the relevant matrix [30, 31].

Equipment
High-pressure microwave digestion apparatus "Multiwave" (from Perkin-Elmer, Germany, A. Paar KG, Austria)

Chemicals
Nitric acid, Suprapur (or equivalent) or subboiling distilled

Digestion of blood
(suitable for atomic or mass spectrometric detection. When voltammetric techniques are used, the digestion parameters or the acid combinations have to be modified as required for the PMD device)
Then 1 mL blood and 2 mL HNO_3 are pipetted into 20 mL quartz vessels (accessory). The vessels are sealed as described in the manufacturer's instructions, and inserted into the pressure vessels and the microwave oven.

	Step	Power	Duration	Final power
Digestion program:	1	100 W	5 minutes	500 W
	2	1000 W	15 minutes	1000 W
	3	0 W	15 minutes	0 W

Then the clear digestion solution is diluted to the desired sample volume with H_2O.

Interference and discussion
The same statements apply to this device as to the PMD device. The analysis time for 6 samples, including the cooling phase, is approx. 2 h, so that about 20 samples can also be ashed per day.

7.1.10 Example 10: High-pressure digestion using the "High Pressure Asher (HPA)"

This apparatus consists of an autoclave with conventional heating and quartz glass sample vessels.

Equipment
High-pressure digestion apparatus "HPA" (from Perkin-Elmer, Germany) or "HPAS" (A. Paar KG, Austria)

These devices allow pressures up to 130 bar and a temperature up to 320 °C for digestion, and they therefore achieve, together with the appropriate acid mixtures, a very high oxidation potential [18].

Chemicals
Nitric acid, Suprapur (or equivalent) or subboiling distilled
Perchloric acid, Suprapur
Nitrogen 200 bar

Digestion of blood (atomic or mass spectrometric detection)
1 mL blood and 2 mL HNO_3 are pipetted into quartz glass vessels (30 mL), which are sealed in accordance with the manufacturer's instructions and inserted into the pressure vessel.
The following digestion program is sufficient:

Step	Start temp.	Time	End temp.
1	70 °C	20 min	120 °C
2	150 °C	60 min	200 °C
3	280 °C	70 min	280 °C

The solution is filled to the desired final volume with ultrapure water.

Digestion of blood (voltammetric detection)
1 mL blood, 2 mL HNO_3 and 0.25 mL $HClO_4$ are pipetted into the digestion vessel.
The following digestion program is sufficient:

Step	Start temp.	Time	End temp.
1	70 °C	20 min	120 °C
2	150 °C	60 min	200 °C
3	300 °C	70 min	300 °C

The HNO_3 is evaporated in a quartz crucible and the residue is taken up with $HClO_4$ and H_2O and filled to the desired final volume with ultrapure water.

Digestion of serum for determination of iodine
1 mL serum and 2.3 mL acid mixture (mixture: 10 mL HNO_3 + 2.5 mL $HClO_4$) are pipetted into the digestion vessel.
The following digestion program is sufficient:

Step	Start temp.	Time	End temp.
1	70 °C	20 min	120 °C
2	150 °C	60 min	200 °C
3	280 °C	70 min	280 °C

The solution is filled to the desired final volume with ultrapure water.
The same parameters can be used for urine as for blood or serum.

Interference and discussion
This is a highly efficient version of a pressure digestion system on account of the elevated temperature and the high pressure. No interference to subsequent analytical procedures has been observed.

7 Appendix

7.1.11 Example 11: Digestion of liquids in a high-pressure microwave flowthrough digestion apparatus

This is a flowthrough system heated by microwaves under a pressure of approx. 30 bar, which permits continuous digestion of up to 60 liquid samples (e.g. blood, serum or urine) via a suitable autosampler. Therefore on-line coupling to analytical procedures such as ICP-AES or ICP-MS is possible.

Equipment
High-pressure microwave flowthrough digestion apparatus "Autoflow" (Berghoff, Germany, at present being developed, not yet commercially available)
This device is connected to an autosampler for samples and digestion solutions so that continuous and automated operation is ensured [23].

Chemicals
Nitric acid, Suprapur (or equivalent) or subboiling distilled
Digestion of serum or milk (for atomic or mass spectrometric detection)
The following parameters are set via the computer:

Loop: 5 mL
Dwell time: 3 minutes
Pressure: 25 bar
Power: 100 W
Temperature: 234 °C

As this device is not yet commercially available, changes to the parameters are still possible (it is also intended to include on-line coupling to measurement systems).

7.1.12 Example 12: UV digestion of liquids

The organic matrix is destroyed by the OH radicals formed from H_2O_2 added to the sample under the influence of UV radiation. This digestion is especially suitable for aqueous and urine samples, but can also be used for serum. Twelve samples can be digested simultaneously using the apparatus described here.

Equipment
UV digestion device "UV 1000" (from Kürner Analysentechnik, Germany)
The reactor can be loaded with up to 12 samples in one run.

Chemicals
Nitric acid, Suprapur (or equivalent) or subboiling distilled
Hydrogen peroxide

Digestion of serum or urine
5 mL of the sample and 2 mL HNO_3 are pipetted into the sample vessels, which are placed in the reactor according to the manufacturer's instructions.
The reaction proceeds according to the following schedule:

After 1 h addition of 1 mL H_2O_2
After 3 h addition of 1 mL H_2O_2
Total duration 5 h

In the case of urine no addition of HNO_3 is necessary. In this case 5 mL H_2O_2 can be added directly to 5 mL urine. Thus the time required is reduced to approximately 1 hour.

Interference and discussion

This procedure, which was mainly developed for ashing organic constituents in aqueous systems, also permits the satisfactory digestion of complex matrices. Reagents can be added without difficulty using an Eppendorf pipette (or equivalent) via apertures in the UV protective glass, whereby the device remains in operation [24, 25].
The procedure is especially suitable for atomic and mass spectrometric analytical methods. The digestion efficiency is not sufficiently satisfactory for voltammetry.

8 References

[1] *R. Bock:* A Handbook of Decomposition Methods in Analytical Chemistry. Internat. Textbook Comp. Ltd., London (1979)

[2] *P. Schramel, I. Wendler* and *G. Knapp:* Total digestion of silicate containing matrices (plant, soil, sludges) using a pressure ashing device with PFA-vessels. Fresenius J. Anal. Chem. 356, 512–514 (1996)

[3] *S. Hasse* and *P. Schramel:* Voltammetric Determination of Cd, Cu, Co, Ni and Pb in Milkpowder and Other Biological Materials with Special Regard to the Ashing Method. Mikrochimica Acta III, 449–455 (1983)

[4] *P. Schramel* and *S. Hasse:* Destruction of organic materials by pressurized microwave digestion. Fresenius J. Anal. Chem. 346, 794–799 (1993)

[5] *P. Schramel* and *S. Hasse:* Anwendung voltammetrischer Verfahren zur Spurenelementbestimmung in klinischen Proben: „Eine kritische Bewertung der Möglichkeiten und Grenzen". Lab. med. 17, 491–495 (1993)

[6] *P. Schramel, Li-Qiang Xu, G. Knapp* and *M. Michaelis:* Multi-elemental analysis in biological samples by on-line preconcentration on 8-hydroxyquinoline-cellulose microcolumn coupled to simultaneous ICP-AES. Fresenius J. Anal. Chem. 345, 600–606 (1993)

[7] *E. Scheubeck, A. Nielsen* and *G. Iwantscheff:* Schnellverfahren für den Aufschluss von größeren Mengen an Biomaterial zur analytischen Erfassung von Schwermetallspuren. Fresenius Z. Anal. Chem. 294, 398–401 (1979)

[8] *E. Scheubeck, J. Gehring* and *M. Pickel:* Eine Druckaufschlusseinrichtung für die schnelle Aufbereitung von größeren Substanzmengen an Biomaterialien und organischen Substanzen zur analytischen Erfassung von Schwermetallspuren. Fresenius Z. Anal. Chem. 297, 113–116 (1979)

[9] *P. Schramel* and *S. Hasse:* Iodine Determination in Biological Materials by ICP-MS. Mikrochim. Acta 166, 205–209 (1994)

[10] *R. Bock:* Aufschlussmethoden der anorganischen und organischen Chemie. Verlag Chemie, Weinheim (1978)

[11] *L. Klotz, G. Kaiser, P. Tschöpel* and *G. Tölg:* Aufschluss biologischer Matrices für die Bestimmung sehr niedriger Spurengehalte bei begrenzter Einwaage mit Salpetersäure unter Druck in einem Teflongefäß. Fresenius Z. Anal. Chem. 260, 207 (1972)

[12] *B. Bernas*: New method for decomposition and comprehensive analysis of silicates by atomic absorption spectrometry Anal. Chem. 40, 1682 (1968)

[13] *F. J. Langmyhr* and *P. E. Paus.* Anal. Chim. Acta 49, 358 (1949)

8 References

[14] *M. Stoeppler* and *F. Backhaus:* Pretreatment Studies with Biological and Environmental Materials. I. Systems for Pressurized Multisample Decomposition. Fresenius Z. Anal. Chem. 291, 116–120 (1978)

[15] *L. Klotz, G. Henze, G. Kaiser, S. Pahlke, M. Veber* and *G. Tölg*: Wet Mineralization of Organic Matrices in Glassy Carbon Vessels in a Pressure-Bomb System for Trace Element Analysis. Talanta 26, 681–691 (1979)

[16] *F. Backhaus, M. Cremer, H.J. Groteklaes, A. Hertling, O. Schneider* and *M. Stoeppler:* Entwicklung und Erprobung von Druckaufschlußsystemen variabler Kapazität. Ber. d. KFA Jülich (1981)

[17] *P. Schramel, A. Wolf, R. Seif* and *B.-J. Klose:* Eine neue Apparatur zur Druckveraschung von biologischem Material. Fresenius Z. Anal. Chem. 302, 62–64 (1980)

[18] *P. Schramel, S. Hasse* and *G. Knapp:* Einsatz des Hochdruckveraschers HPA nach Knapp [1] für die voltammetrische Bestimmung von Spurenelementen in biologischem Material. Fresenius Z. Anal. Chem. 326, 142–145 (1987)

[19] *P. Schramel* and *S. Hasse:* Destruction of organic materials by pressurized microwave digestion. Fresenius J. Anal. Chem. 34, 794–799 (1993)

[20] *S. Hasse* and *P. Schramel:* Voltammetric Determination of Cd, Cu, Co, Ni and Pb in Milkpowder and Other Biological Materials with Special Regard to the Ashing Method. Mikrochimica Acta (Wien) III, 449–455 (1983)

[21] *P. Schramel* and *Li-Quiang Xu:* On-line preconcentration of arsenic from water samples with an anion-exchanger column coupled to a hydride generation ICP system. Fresenius J. Anal. Chem. 343, 373–377 (1992)

[22] *M. Würfels, E. Jackwerth* and *M. Stoeppler:* Probenvorbehandlungsstudien mit biologischen und Umweltmaterialien. V. Zum Problem der Störung inversvoltammetrischer Spurenanalysen nach Druckaufschluss biologischer Probenmaterialien. Fresenius Z. Anal. Chem. 329, 459–461 (1987)

[23] *U. Pichler, G. Knapp* and *M. Michaelis:* Microwave-heated flow-through system for high temperatures. In: Canas 95, 337–340 (1996)

[24] *J. Golimowski* and *K. Golimowska:* UV-photooxidation as pretreatment step in inorganic analysis of environmental samples. Analytica Chimica Acta 325, 111–133 (1996)

[25] Erfahrungsbericht UV-Aufschluss. Inst. f. Anorganische Chemie der TU Clausthal, Clausthal-Zellerfeld

[26] *M. Stoeppler, K.P. Müller* and *F. Backhaus:* Pretreatment Studies with Biological and Environmental Materials. III. Pressure Evaluation and Carbon Balance in Pressurized Decomposition with Nitric Acid. Fresenius Z. Anal. Chem. 297, 107–112 (1979)

[27] *E. Schwarzbach:* Schneller Aufschluss von Blut mit Wasserstoffperoxid. Arbeitsmed. Sozialmed. Arbeitshyg. 6, 45–46 (1971)

[28] *H. Hein:* Bleibestimmung in Blut mit der Graphitrohrküvette HGA-72. Analysentechn. Ber. 29, Bodenseewerk Perkin Elmer u. Co. GmbH (1974)

[29] *W.R. Hatch* and *W.L. Ott:* Determination of Sub-Microgram Quantities of Mercury by Atomic Absorption Spectrophotometry. Anal. Chem. 40, 2085–2087 (1968)

[30] *M. Zischka, P. Kettisch, A. Schalk* and *P. Kainrath:* New developments in high-pressure wet-ashing procedures. Laborpraxis, 34–41 (1995)

[31] *M. Zischka, G. Knapp, P. Kettisch* and *A. Schalk:* Microwave pressure digestion – development, performance and safety of a novel system of sample preparation. Canas '95, 355–359 (1995)

Author: *P. Schramel*

N-Benzylvaline after exposure to benzylchloride

Application Determination of the globin adduct in blood

Analytical principle Capillary gas chromatography/
mass spectrometric detection (MS)

Completed in May 2001

Summary

The procedure described here permits determination of the adduct of benzylchloride on the N-terminal valine of human haemoglobin.

For this purpose erythrocytes are separated from EDTA-stabilised whole blood. After isolation of globin from the erythrocytes, the N-terminal benzylvaline adduct is cleaved from the globin in the presence of formamide by means of a modified Edman degradation. After capillary gas chromatographic separation, the measurement is carried out in a mass selective detector in the SIM mode.

Calibration graphs are plotted with calibration standards prepared with pooled globin for quantitative evaluation. The calibration standards are treated in the same manner as the globin samples to be investigated. A commercially available dipeptide standard (N-benzyl-valine-leucine-anilide) is used for calibration. The dipeptide N-2-ethoxyethyl-valine-alanine-anilide serves as the internal standard.

N-Benzylvaline

Within-series imprecision: Standard deviation (rel.) $s = 19.3\%$
Prognostic range (rel.) $u = 43.0\%$
at an N-benzylvaline concentration of 11.8 µg/L
and where $n = 10$ determinations

Between-day imprecision: Standard deviation (rel.) $s = 20.5\%$
Prognostic range (rel.) $u = 45.7\%$
at an N-benzylvaline concentration of 47.2 µg/L
and where $n = 10$ determinations

Quantitation limit: 0.5 µg/L blood

Benzylchloride

Benzylchloride (CAS No.: 100-44-7) is a colourless liquid that gives off fumes in humid air and has a pungent odour. The odour threshold is in the range of 0.05 to 2 mL/m^3 [1–3].

Benzylchloride is a typical intermediate product in the chemical industry and is further processed, for example to benzyl alcohol, phthalic acid benzyl esters (benzyl phthalates as plasticisers), benzylamine and derivatives (intermediate products for pigments) as well as other precursors (e.g. auxiliary materials for the textile industry, catalysts, emulsifiers, disinfectants). There is no known direct use of benzylchloride as a finished product [1, 4–8].

Due to the relatively high vapour pressure of this lipophilic substance exposure can occur in a variety of circumstances during handling of the intermediate product in the industrial environment, especially when sampling, distilling and filling liquid benzylchloride [1, 4].

Benzylchloride causes marked irritation to the mucous membranes and conjunctiva. Concentrations in the ambient air that induce irritation of the mucous membranes can lead to liver dysfunction and neurological symptoms. Benzylchloride has exhibited a direct mutagenic effect in various test systems. In animal studies benzylchloride proved carcinogenic after subcutaneous, epicutaneous and intragastral administration of the substance to rats and mice. Both local and systemic tumours were induced as a result of intestinal intake. The Commission for the Investigation of Health Hazards of Chemical Compounds in the Work Area assigned benzylchloride to category 2 of the carcinogenic working materials in 1992 [9, 10].

The MAK value was annulled in 1992 [9, 10]. The American TLV value is 1 ppm [11]. In 1987 benzylchloride was placed in carcinogen class C in Sweden, and it was assigned a TWA value of 5 µg/m^3 [12].

At present the important details of the metabolism of benzylchloride in humans, which are necessary for reliable biomonitoring, are not available. Animal studies have shown that benzylchloride is rapidly metabolised after absorption and is excreted, predominantly via the kidneys, after a short time. More than 65% of the dose is excreted within 24 hours [10, 13].

The toxic irritating effects of benzylchloride may be caused by hydrochloric acid, which is formed during the spontaneous or enzymatically catalysed hydrolysis of benzylchloride [13]:

$$\text{C}_6\text{H}_5\text{-CH}_2\text{-Cl} + \text{GSH} \xrightarrow{\text{(GST)}} \text{C}_6\text{H}_5\text{-CH}_2\text{-SG} + \text{HCl}$$

The glutathione adduct formed as an intermediate in the metabolism of benzylchloride can penetrate the erythrocyte membranes and thus initiate conjugation with the N-terminal valine of the globin (Figure 1) [14].

The main metabolite was found to be benzyl mercapturic acid, but hippuric acid and benzyl cysteine were also present as by-products. The oxidative degradation of benzylchloride to the substituted benzoic acid (or the subsequent biotransformation to

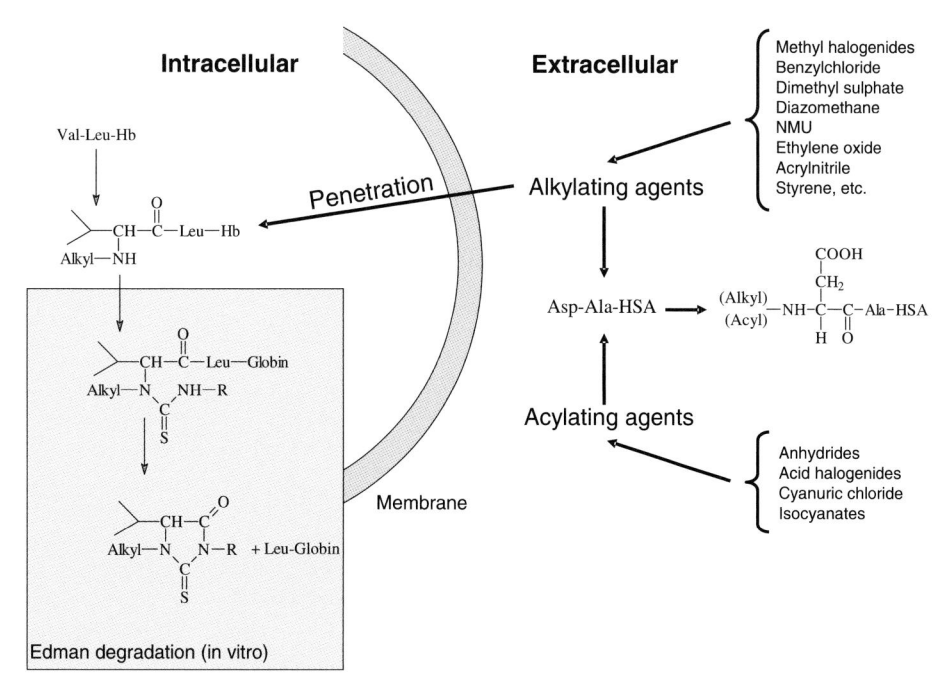

Fig. 1. Extracellular and intracellular alkylation and acylation by reactive substrates (R = isothiocyanate group; Hb = haemoglobin; HSA = human serum albumin)

hippuric acid) proceeds to a significantly less extent than in the case of toluene or o-nitrotoluene [10, 13].

As a consequence of the assignment of benzylchloride to category 2 of the carcinogenic substances, workers who handle benzylchloride must undergo a regular examination as part of the preventive measures implemented in occupational medicine in accordance with G40 issued by the BG-Chemie (Chemical Employers Mutual Indemnity Association in Germany) [15].

It is necessary to monitor and document the levels of inner benzylchloride exposure that may occur despite occupational health and safety precautions. The analytical procedure presented here determines the "formally genotoxic fraction" of the benzylchloride metabolism in the Hb adduct, which can in principle also lead to DNA conjugation. No adverse effects were observed in persons who had been handling benzylchloride for more than 20 years at levels below the current empirical value acceptable to occupational medicine (arbeitsmedizinischer Erfahrungswert, AEW) of 10 µg N-benzylvaline (released from the globin adduct) [16].

Authors: *J. Lewalter, G. Leng, D. Ellrich*
Examiner: *J. Angerer*

N-Benzylvaline after exposure to benzylchloride

Application   Determination of the globin adduct in blood

Analytical principle   Capillary gas chromatography/
mass spectrometric detection (MS)

Completed in   May 2001

Contents

1   General principles
2   Equipment, chemicals and solutions
2.1 Equipment
2.2 Chemicals
2.3 Solutions
3   Specimen collection and sample preparation
3.1 Sample preparation
4   Operational parameters
4.1 Operational parameters for gas chromatography and mass spectrometry
5   Analytical determination
6   Calibration
7   Calculation of the analytical result
8   Standardisation and quality control
9   Reliability of the method
9.1 Precision
9.2 Accuracy
9.3 Detection limits
9.4 Sources of error
10   Discussion of the method
11   References

1 General principles

Erythrocytes are separated from EDTA-stabilised whole blood. After isolation of globin from the erythrocytes, the N-terminal benzylvaline adduct is cleaved from the globin in the presence of formamide by means of a modified Edman degradation.

After capillary gas chromatographic separation, the measurement is carried out in a mass selective detector in the SIM mode.

Calibration graphs are plotted with calibration standards prepared with pooled globin for quantitative evaluation. The calibration standards are treated in the same manner as the globin samples to be investigated. A commercially available dipeptide standard (N-benzyl-val-leu-anilide) is used for calibration. The dipeptide N-2-ethoxy-ethyl-valine-alanine-anilide serves as the internal standard.

2 Equipment, chemicals and solutions

2.1 Equipment

Gas chromatograph/mass spectrometer with the possibility of electronic pressure programming and of selected ion detection and an integration system or a data processing system

Capillary column DB-5-MS (5%-phenyl-methylpolysiloxane), length 50 m; inner diameter 0.2 mm; film thickness 0.33 µm (e.g. from J&W)

Volumetric flasks 10 mL, 100 mL and 500 mL

Centrifuge (e.g. Hettich, Rotanta/RP)

Nitrogen station (e.g. from Zymark)

4 mL Sample vials (e.g. from Sarstedt)

10 mL Duran screw-top tubes 100 mm×16 mm×1.2 mm (e.g. from Schütt)

50 mL Polyethylene tubes (e.g. Falcon tubes from Becton Dickinson)

0.2 mL Microvials (e.g. from Macherey-Nagel)

Laboratory shaker (e.g. Vibrax from Ika)

Water bath with shaking device (e.g. from Julabo)

10 µL, 100 µL, 1000 µL Pipettes (e.g. from Eppendorf)

Analytical balance (e.g. from Mettler)

Desiccator

2.2 Chemicals

Unless otherwise specified, all the chemicals used must be at least p.a. grade

Bidistilled water (equivalent to ASTM Type I)

Formamide (e.g. from Calbiochem)

1 M NaOH (e.g. from B. Kraft)

1 M Hydrochloric acid (e.g. from Merck)

Sodium carbonate p.a. (e.g. from Merck)

Pentafluorophenylisothiocyanate (e.g. from Fluka)

tert-Butylmethyl ether for organic trace analysis (e.g. Suprasolv® from Merck)

Toluene for organic trace analysis (e.g. Suprasolv® from Merck)

N-Benzyl-valine-leucine-anilide (e.g. from Bachem)

N-2-Ethoxyethyl-valine-alanine-anilide (e.g. from Bachem)

Ethanol (e.g. from Sigma)

2-Propanol (e.g. from Riedel-de Haën)

Ethyl acetate (e.g. from Merck)

pH test strips (pH range 7 to 11, e.g. from Macherey-Nagel)

2.3 Solutions

0.1 M Sodium carbonate solution:

5.3 g sodium carbonate are weighed into a 500 mL volumetric flask. The flask is subsequently filled to its nominal volume with bidistilled water.

Solutions of the internal standard

Starting solution:
Approximately 17.7 mg 2-ethoxyethyl-valine-analine-analide are weighed exactly into a 100 mL volumetric flask. The flask is subsequently filled to its nominal volume with ethanol (100 mg/L).

Stock solution:
100 µL of the starting solution of the internal standard are pipetted into a 10 mL volumetric flask. The flask is subsequently filled to its nominal volume with bidistilled water (1 mg/L).

Formamide:
Any NH_3 impurities present in formamide are purged from the substance by passing a stream of nitrogen through the formamide until the pH value is approx. 8 in the stream of gas above the flask. The pH value is checked using pH test strips.

HCl/propanol mixture for the isolation of globin:
50 mL of 1 M HCl are placed in a 1 L volumetric flask. The volumetric flask is then filled to its nominal volume with 2-propanol.

2.4 Calibration standards

Calibration standards and control material are prepared by spiking either pooled globin from test subjects without any known exposure to benzylchloride or by spiking commercially available globin. Globin from several test subjects is isolated in accordance with Section 3.1 and collected in a sealable container in order to prepare the pooled globin. The globin is thoroughly mixed on a sample shaker for at least 12 h and stored at approx. −18 °C until the standards and the control material are to be prepared.

All starting and stock solutions are stored in sealed vessels at approx. +4 °C, and they are stable under these conditions for at least 2 weeks.

Starting solution:
Approximately 19.1 mg N-benzyl-valine-leucine-anilide are weighed exactly into a 100 mL volumetric flask. The flask is then filled to its nominal volume with ethanol (100 mg/L).

Stock solution A:
1 mL of the starting solution is pipetted into a 10 mL volumetric flask. The flask is filled to its nominal volume with bidistilled water (10 mg/L).

Stock solution B:
100 µL of the starting solution are pipetted into a 10 mL volumetric flask. The flask is filled to its nominal volume with bidistilled water (1 mg/L).

Stock solution C:
10 µL of the starting solution are pipetted into a 10 mL volumetric flask. The flask is then filled to its nominal volume with bidistilled water (0.1 mg/L).

Calibration standards are prepared in pooled globin from these stock solutions before each analytical run. For this purpose approx. 100 mg globin are weighed exactly on an analytical balance into a Duran screw-top tube, and 1.5 mL formamide are added. The sample is shaken (approx. 15 min) on a shaker (e.g. Ika Vibrax) until the solution is clear. Then 100 µL of the stock solution of the internal standard and the appropriate volumes of stock solutions A, B or C are pipetted according to the following pipetting scheme (Table 1). The calibration standards prepared as described are treated in the same manner as the samples to be investigated (see Section 3.1). The error resulting from dilution is negligible.

Table 1. Pipetting scheme for the preparation of the calibration standards

N-Benzylvaline in		Calibration standards			Internal standard
(µg/L blood)*	(pmol/g globin)	Stock solution A [µL]	Stock solution B [µL]	Stock solution C [µL]	[µL]
BV	BV	–	–	–	100
1.6	50	–	–	10	100
3.1	97	–	–	20	100
4.7	146	–	–	30	100
11.8	360	–	–	75	100
19.7	611	–	–	125	100
47.2	1463	–	30	–	100
118.0	3658	–	75	–	100
197.0	6107	–	125	–	100
393.0	12183	25	–	–	100

BV = Blank value
* The concentration in µg/L blood must be regarded as an approximate value. The calculation is based on the assumption that human blood has a mean content of 158.5 g globin per litre.

3 Specimen collection and sample preparation

After blood samples have been taken in EDTA Monovettes, the isolation of globin should be carried out as rapidly as possible. However, blood samples can be stored in the refrigerator for 3 days at 4 to 6 °C before the globin is isolated.

Isolation of erythrocytes
5 mL samples of EDTA blood are centrifuged for 5 minutes at 1200 g. The plasma is drawn off using a pipette and discarded.
2.5 mL of the 0.9% (154 mmol) sodium chloride solution are added to the erythrocyte sediment, and they are mixed thoroughly. Centrifugation for 5 minutes at 1200 g is repeated. The supernatant is drawn off using a pipette and discarded. This washing step is repeated until the supernatant is colourless (at least twice). Approximately 2.5 mL of erythrocyte sediment is obtained from 5 mL blood.
To achieve lysis, 2.5 mL bidistilled water are added to the isolated and washed erythrocytes and they are deep-frozen at approx. –20 °C for at least 8 hours.

Isolation of globin
The frozen lysate sample is thawed and then homogenised by shaking. Then the entire lysate is transferred to a 50 mL Falcon tube using a pipette, 30 mL of the HCl/propanol mixture are added, and the sample is shaken vigorously. Then the sample is centrifuged for 10 minutes at 2000 g to remove the cell detritus. The supernatant is transferred into a new 50 mL Falcon tube and the globin is precipitated by adding 20 mL ethyl acetate. After waiting for a period of 15 minutes, the globin is shaken briefly and then the tube is centrifuged for 5 minutes at 1500 g. The supernatant is decanted and discarded. The globin precipitate is subsequently washed four times with approx. 10 mL ethyl acetate in each case, centrifuged each time for 5 minutes

at 1500 g, and the supernatant is decanted and discarded each time. Finally the isolated globin is placed overnight in a desiccator, that has been previously flushed with a large quantity of nitrogen, and is dried under a vacuum.

Storage of the globin sample
Isolated globin in polypropylene tubes (Falcon 2006) can be stored for up to 2 months at +4 °C and up to 12 months in the deep-freezer at -30 °C.

3.1 Sample preparation

Approx. 100 mg globin are weighed exactly into a Duran screw-top tube on an analytical balance, and 1.5 mL formamide (which can contain NH_3, see Section 2.3) are added. The sample is shaken on a shaker (e.g. Ika Vibrax) until the solution is clear (approx. 15 min). Then 100 µL of the stock solution of the internal standard, 30 µL 1 N NaOH and 20 µL pentafluorophenylisothiocyanate are added. The solution is then allowed to react by incubation for 3 hours at 80 °C in a water bath that is being gently shaken.

After cooling, the samples are saturated with NaCl (approx. 1.8 g). Then they are extracted twice, each time with 3.0 mL tertiary butylmethyl ether for 2 minutes on the laboratory shaker. The samples are centrifuged for 5 minutes at 3000 g to improve phase separation. The combined ether phases are transferred to a test-tube and evaporated to dryness at 35 °C in a stream of nitrogen at a nitrogen station. The residue is dissolved in 500 µL toluene and then transferred to a test-tube.

The toluene phase is now washed with 2 mL bidistilled water for approx. 5 minutes on the shaker. After centrifugation (5 minutes at 3000 g), the lower aqueous phase is withdrawn and discarded. 2 mL of freshly prepared 0.1 M Na_2CO_3 solution are added and the sample is extracted for approx. 5 minutes on the shaker in a second washing step. After centrifugation, the upper toluene phase is removed without water and transferred to a sample vial (4 mL). The sample is evaporated to dryness under a stream of nitrogen at a nitrogen station. The anhydrous residue is finally dissolved in 50 µL toluene, transferred to microvials and sealed tightly.

Storage of the processed samples
The samples thus processed can be stored at +4 °C for 4 weeks or at –30 °C for 6 months.

4 Operational parameters

4.1 Operational parameters for gas chromatography and mass spectrometry

Capillary column:	Material:	Fused silica
	Stationary phase:	DB-5-MS
	Length:	50 m

	Inner diameter:	0.2 mm
	Film thickness:	0.33 µm
Detector:	Mass selective detector (MSD)	
Temperatures:	Column:	Initial temperature 120 °C for 2 minutes, then increase at a rate of 15 °C/min to 220 °C, then at a rate of 20 °C/min to 280 °C, then 20 min at the final temperature
	Injector:	280 °C
	Transfer line:	300 °C
Carrier gas:	Helium 5.0 with a constant flow of 0.7 mL per minute	
Split:	Splitless, split on after 1 minute at 57 mL/min	
Septum purging:	3 mL per minute	
Sample volume:	0.5 µL (pulsed splitless, 3 bar for 2 minutes)	
Ionisation type:	Electron impact ionisation (EI)	
Ionisation energy:	70 eV	
Dwell time:	200 ms	
Electron multiplier:	Relative +400 V	

All other parameters must be optimised in accordance with the manufacturer's instructions.

5 Analytical determination

In order to carry out the determination, 0.5 µL of each of the globin samples processed as described in Section 3.1 is injected into the gas chromatograph operating in the pulsed splitless mode.

If the measured values are above the linear range of the calibration graphs (>400 µg/L), the globin samples must be diluted with formamide in the ratio of 1:10, processed and measured anew. A quality control sample is analysed with each analytical series (see Section 8).

The temporal profiles of the ion fragments shown in Table 2 are recorded in the SIM mode:

Table 2. Retention times and SIM ion fragments

Compound	Retention time [min]	Masses
Benzylvaline-F5-PTH	23.51	297
		414*
N-2-Ethoxyethylvaline-F5-PTH (IS)	19.14	396*
		363

The masses marked with * are used for quantitative evaluation
(F5-PTH = Pentafluorophenylthiohydantoin)

The retention times shown in Table 2 serve only as a guide. Users of the method must satisfy themselves of the separation power of the capillary column used and the resulting retention behaviour of the substances. Figure 2 shows an example of a chromatogram of a sample from a person who was not exposed to the analyte, and Figure 3 shows the chromatogram of a calibration standard with a benzylvaline adduct concentration of 15.8 µg/L. In the chromatogram shown as an example in Figure 4 the processed globin sample of a person was found to contain a concentration of 1.9 µg benzylchloride (released from the globin adduct) per litre blood.

6 Calibration

The globin calibration standards are processed in the same manner as the globin samples (Section 3.1) and analysed by means of gas chromatography/mass spectrometry as described in Sections 4 and 5. Linear calibration graphs are obtained by plotting the quotients of the peak areas of the benzylvaline derivative and that of the internal standard as a function of the concentrations used. It is unnecessary to plot a complete calibration graph for every analytical series. It is sufficient to analyse one calibration standard for every analytical series. The ratio of the result obtained for this standard and the result for the equivalent standard in the complete calibration graph is calculated. Using this quotient, each result read off the calibration graph is adjusted for the relevant series. A typical linear calibration graph is shown in Figure 5. The qualifier has an intensity of approx. 76% with respect to the quantifier under the conditions given here.
New calibration graphs should be plotted if the quality control results indicate systematic deviation.
The calibration graph is linear between the detection limit and 400 µg per litre blood.

7 Calculation of the analytical result

Quotients are calculated by dividing the peak areas of the analyte by the peak area of the internal standard. These quotients are used to read off the pertinent concentration of benzylvaline adduct in µg per litre from the relevant calibration graph.

8 Standardisation and quality control

Quality control of the analytical results is carried out as stipulated in the guidelines of the Bundesärztekammer (German Medical Association) [17, 18] and in the special preliminary remarks to this series. In order to determine the precision of the method, a globin control sample containing a constant concentration of the benzylvaline dipeptide is analysed. As material for quality control is not commercially available, it must be prepared in the laboratory. A defined amount of the benzylvaline dipeptide is added to pooled globin (e.g. from Sigma) for this purpose. The concentration of this control material should lie within the decisive concentration range. The control material must be freshly prepared before each analytical series. The theoretical value and the tolerance range for this quality control material are determined in the course of a pre-analytical period (one analysis of the control material on each of 20 different days) [19, 20].

9 Reliability of the method

9.1 Precision

Quality control samples spiked with 11.8 µg/L and 47.2 µg/L of the benzylvaline dipeptide were processed and analysed to check the precision in the series. When these globin samples were determined 10 times, the relative standard deviations were 19.3% for 11.8 µg/L and 18.5% for 47.2 µg/L, equivalent to a prognostic range of 43.0% and 41.2% respectively (cf. Table 3).

Table 3. Precision for the determination of the benzylvaline adduct

	n	Concentration [µg/L]	Standard deviation (rel.) [%]	Prognostic range [%]
In the series	10	11.8	19.3	43.0
	10	47.2	18.5	41.2
From day to day	10	50.0	20.5	45.7

In addition, the precision from day to day was determined. The same material (concentration = 50.0 µg/L) used for the determination of the precision in the series was analysed in this case. These globin solutions were processed and analysed on 10 different days. The standard deviation was 20.5%, which is equivalent to a prognostic range of 45.7% (cf. Table 3).

9.2 Accuracy

Recovery experiments were performed to check the accuracy of the method. For this purpose globin without the analyte and spiked with defined quantities of the benzylvaline dipeptide was analysed 10 times. The relative recovery rates were 96% at a concentration of 11.8 µg/L and 99% at 47.2 µg/L. The accuracy of the method was not determined by measuring the losses due to processing, as no appropriate reference substances were available.

9.3 Detection limits

A detection limit of 0.5 µg/L blood was calculated as three times the ratio of the analyte signal to the background noise signal.

9.4 Sources of error

As globin cannot be selectively precipitated from haemolysed whole blood, it is important to ensure that intact erythrocytes are separated from the blood plasma. They must then be carefully washed with sodium chloride solution to remove other proteins, especially serum albumin, as otherwise the measurement result that is based on the weighed-in amount of globin is falsified. Erythrocyte haemolysate should not be stored deep-frozen for longer than the given period, as an increase in the adduct quantity was observed after 6 months in individual cases [21]. Therefore the globin should be isolated as rapidly as possible. The adduct content of globin stored at approx. −18°C is then stable for at least 12 months.
The formamide used is of decisive importance for the success of the derivatisation reaction. If the formamide is contaminated with amines or ammonia due to storage conditions or the type of product supplied, this can affect the pH value of the reaction solution and thus lead to reduction in the reaction yield. Contamination of the chemicals used was not observed. Similarly, the author has observed no analytical background interference that adversely affected quantification during analysis of more than 1000 samples.

10 Discussion of the method

Determination of adducts of alkylating substances on the N-terminal valine by means of the modified Edman degradation involves relatively laborious sample preparation. However, many simplifications and improvements have been achieved in the analysis of adducts since it was introduced as a tool for monitoring the biochemical effects of carcinogenic working materials [22]. The method presented here is based on a procedure developed by Törnqvist et al. [23], but it uses the N-benzyl-valine-leucine-anilide dipeptide as a calibration standard and N-2-ethoxyethyl-valine-leucine-anilide as an internal standard [14]. The use of the dipeptides as standards has great benefits for the Edman degradation. A standard with an alkylated valine bound to a further amino acid (leucine) by a peptide bond is used to determine the adduct content. Only in such a case are the pK_a values of the N-terminus of the standard equivalent to those of the globin-bearing N-benzylvaline, and as a consequence comparable derivatisation yields are achieved. Moreover, reproducibility is improved in all the procedural steps through the use of an internal standard. The quality criteria are considered to be good in view of the numerous steps carried out in sample processing. The detection limit is sufficiently low for the field of occupational medicine. No ubiquitous background interference has been observed [14]. The procedure is relatively laborious, but practicable. An experienced laboratory technician is able to carry out about 32 analyses in 3 days. Compared to conventional injection, the use of the pulsed splitless technique for introducing the samples into the GC column resulted in improved peak symmetry and enhanced detection sensitivity.

Instruments used:
HP 6890 Series GC system (Agilent) with HP 5973 MSD and HP 6890 series injector.

11 References

[1] *Berufsgenossenschaft Chemie:* Benzylchlorid, Benzalchlorid, Benzotrichlorid, Merkblatt M046 (1/84) der BG Chemie, Jedermann Verlag, Heidelberg (1984)
[2] *J.E. Amoore* and *E. Hautala:* Odor as an aid to chemical safety: odor thresholds compared with threshold limit values and volatilities for 214 industrial chemicals in air and water dilution. J. Appl. Toxicol. 3 (6), 272–290 (1983)
[3] *G. Leonardos, D. Kendall* and *N. Barnard:* Odor threshold determinations of 53 odorant chemicals. J. Air Poll. Control Assoc. 19, 91–95 (1969)
[4] Gestis-Stoffdatenbank, Benzylchlorid, BIA, Sankt Augustin (1992)
[5] BuA-Stoffbericht 188, Benzylchlorid, Status 1996, S. Hirzel-Verlag, Stuttgart (1996)
[6] *T.V. Mihajlova*: Benzyl chloride. In: *L. Parmeggiani (ed.):* Encyclopedia of Occupational Health and Safety, Internat. Lab. Office, Geneva, 3rd edition, 262 ff. (1983)
[7] *NIOSH:* Niosh Criteria Documents, No. 3, Benzylchloride, U.S. Department of Health and Human Services, Cincinnati (1978)
[8] *M. Rossberg, W. Lendle, G. Pfleiderer, A. Tögel, E.L. Dreher, E.L. Langer, H. Rassaerts, P. Kleinschmidt, H. Strack, U. Beck, K.A. Lipper, T.R. Torkelson, E. Löser* and *K.K. Beutel:* Chlorinated hydrocarbons. In: Ullmann's Encyclopedia of Industrial Chemistry. VCH Verlagsgesellschaft, Weinheim 5th edition, Vol. A6, 355–360 (1986)
[9] *Deutsche Forschungsgemeinschaft*: MAK- und BAT-Werte-Liste 1992, Report 28, Wiley-VCH, Weinheim (1992)
[10] *H. Greim (ed.)*: α-Chlortoluole. Toxikologisch-arbeitsmedizinische Begründung von MAK-Werten, 18th issue. Wiley-VCH, Weinheim (1992)
[11] *ACGIH (American Conference of Governmental Industrial Hygienists)*: Threshold limit values for chemical substances and physical agents and biological exposure indices (1993–1994), Cincinnati (1993)

[12] *L. Holm* and *B. Holmberg:* Exposures to carcinogens and consequences of listing of carcinogens in the Swedish working environment. Regul. Toxicol. Pharmacol. 7, 185–199 (1987)
[13] *Berufsgenossenschaft Chemie:* Benzylchlorid: Toxikologische Bewertung der BG-Chemie, Heidelberg, Nr. 48, Ausgabe 1/97, 37–79 (1996)
[14] *J. Lewalter:* N-Alkylvaline levels in globin as a new type of biomarker in risk assessment of alkylating agents, Int. Arch. Occup. Environ. Health 68, 519–530 (1996)
[15] TRGS 905 (Technische Regeln für Gefahrstoffe): Verzeichnis krebserzeugender, erbgutverändernder oder fortpflanzungsgefährdender Stoffe. Bekanntmachung des BMA vom 15. März 1995 – IIIb4-35125-5 (1995)
[16] *J. Lewalter:* Personal experience in the monitoring of persons exposed to benzylchloride, BASF AG Leverkusen (1980–2001)
[17] *Bundesärztekammer:* Qualitätssicherung der quantitativen Bestimmungen im Laboratorium. Neue Richtlinien der Bundesärztekammer. Dt. Ärztebl. 85, A699–A712 (1988)
[18] *Bundesärztekammer:* Ergänzung der „Richtlinien der Bundesärztekammer zur Qualitätssicherung in medizinischen Laboratorien". Dt. Ärztebl. 91, C159–C161 (1994)
[19] *G. Lehnert, J. Angerer* and *K.H. Schaller:* Statusbericht über die externe Qualitätssicherung arbeits- und umweltmedizinisch-toxikologischer Analysen in biologischen Materialien. Arbeitsmed. Sozialmed. Umweltmed. 33(1), 21–26 (1998)
[20] *J. Angerer* and *G. Lehnert:* Anforderungen an arbeitsmedizinisch-toxikologische Analysen – Stand der Technik. Dt. Ärztebl. 37, C1753–C1760 (1997)
[21] *M. Törnqvist:* Formation of reactive species that lead to hemoglobin adducts during storage of blood samples. Carcinogenesis 11(1), 51–54 (1990)
[22] *P.B. Farmer, H.G. Neumann* and *D. Henschler:* Estimation of exposure of man to substances reacting covalently with macromolecules. Arch. Toxicol. 60(4), 251–260 (1987)
[23] *M. Törnqvist, J. Mowrer, S. Jensen* and *L. Ehrenberg:* Monitoring of environmental cancer initiators through hemoglobin adducts by a modified Edman degradation method. Anal. Biochem. 154(1), 255–266 (1986)

Authors: *J. Lewalter, G. Leng, D. Ellrich*
Examiner: *J. Angerer*

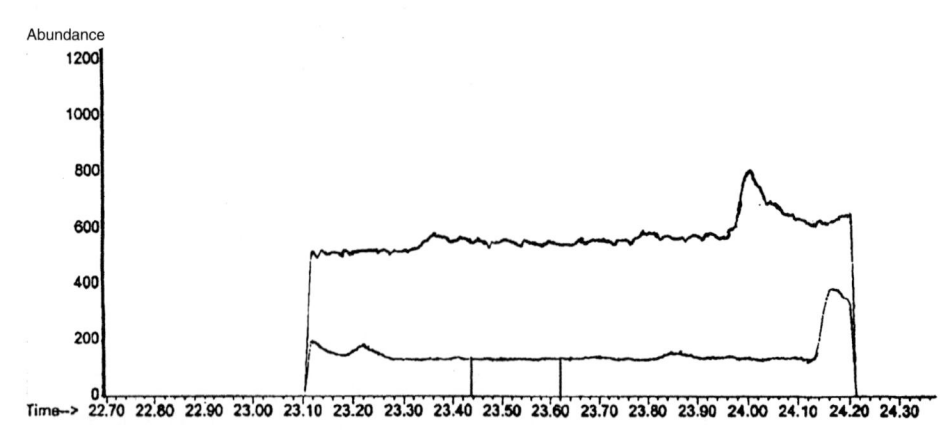

Fig. 2. Example of a chromatogram of a processed globin sample of a person who was not exposed to benzylchloride (Ion 414, Ion 297).

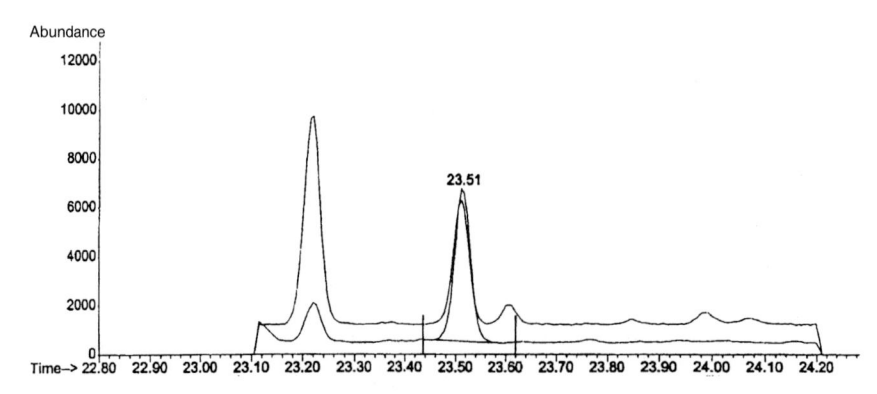

Fig. 3. Example of a chromatogram of a processed globin sample spiked with 15.8 µg per litre blood (Ion 414, Ion 297).

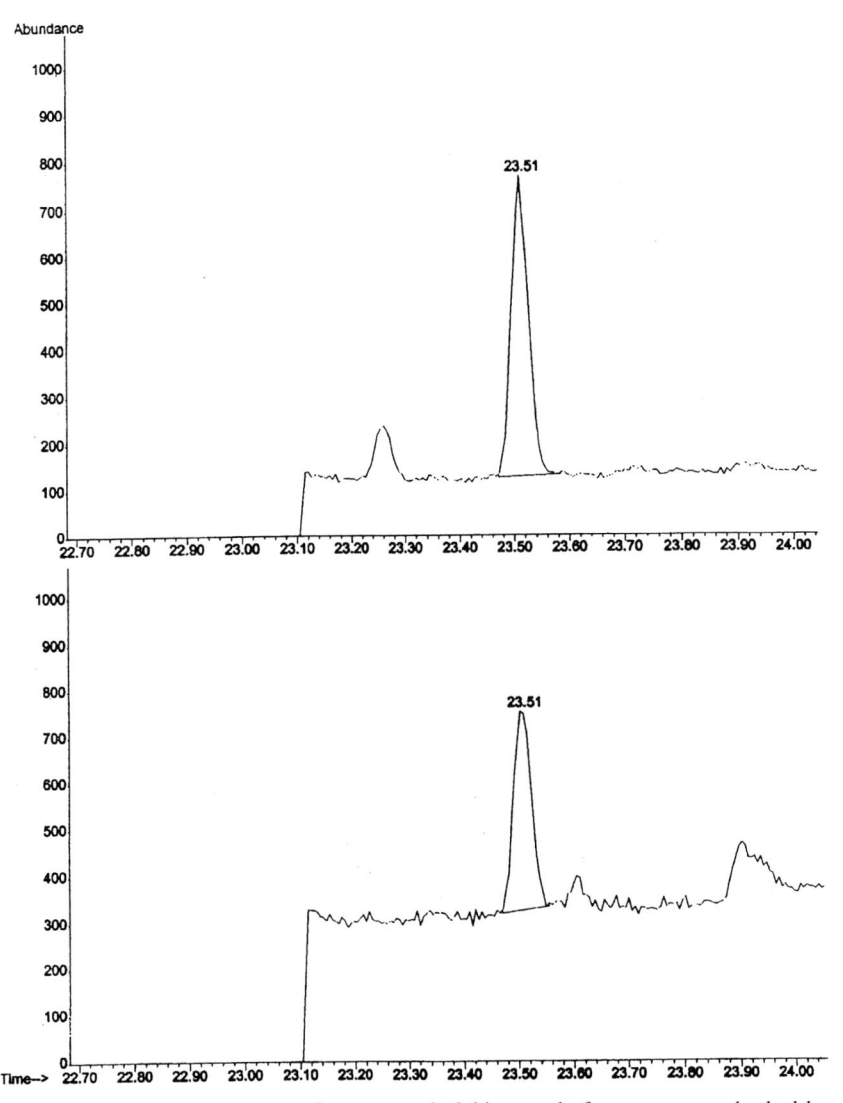

Fig. 4. Example of a chromatogram of a processed globin sample from a person who had been exposed to benzylchloride. The detected concentration of benzylvaline is 1.9 µg/L (Ion 414, Ion 297).

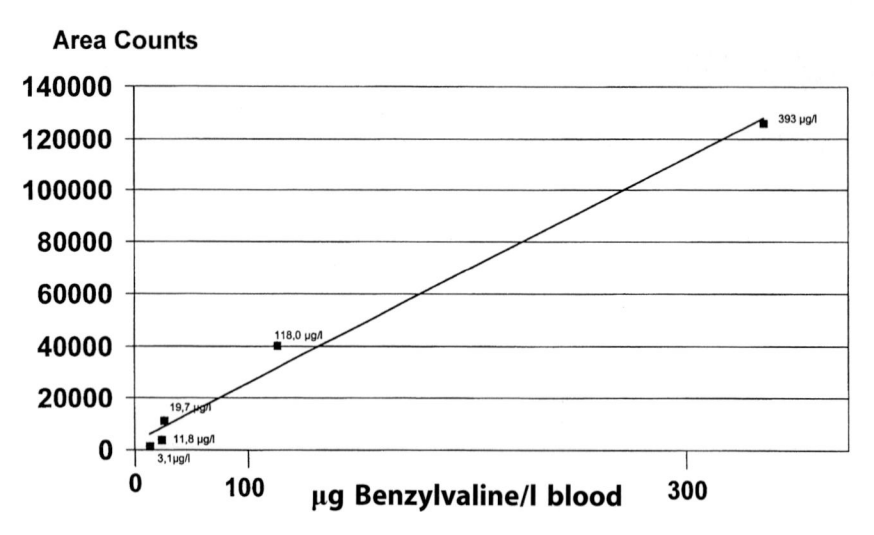

Fig. 5. Calibration function of N-benzylvaline in globin (0–393 µg/L).

Cotinine

Application Determination in urine to ascertain passive exposure to smoking

Analytical principle Capillary gas chromatography/
mass spectrometric detection (MS)

Completed in May 2001

Summary

The method described here serves to quantify cotinine, one of the main metabolites of nicotine excreted in urine. Sensitive determination of cotinine in urine enables differentiation between non-smokers, passive smokers and smokers with a low tobacco consumption.

Deuterated cotinine (cotinine-d_3) is added to a urine sample as an internal standard. Sodium hydroxide is also added, and then extraction is carried out using dichloromethane. The extract is dried under a stream of nitrogen, dissolved in toluene and after capillary gas chromatographic separation, the analyte is quantified by means of mass selective detection in the selective ion monitoring (SIM) mode. Calibration is performed using calibration standards which are prepared in pooled urine and are treated in the same manner as the samples to be analysed.

Cotinine

Within-series imprecision: Standard deviation (rel.) s_w = 6.1% or 1.3%
Prognostic range u = 12.9% or 2.7%
at a concentration of 10 µg or 100 µg cotinine per litre urine
and where n = 17 determinations

Between-day imprecision: Standard deviation (rel.) s_w = 6.4% or 2.8%
Prognostic range u = 13.6% or 5.9%
at a concentration of 10 µg or 100 µg cotinine per litre urine
and where n = 15 determinations

Accuracy: Recovery rate r = 95%

Detection limit: 1 µg cotinine per litre urine

Cotinine

Cotinine is one of the main metabolites of nicotine in the mammalian organism. Determination of cotinine is preferable to determination of nicotine to evaluate a person's smoking status. In contrast to the mother substance, nicotine, exogenous contamination can be virtually excluded in the case of the cotinine metabolite. The cotinine concentration limit between non-smokers who may be passively exposed to smoking and (occasional) smokers is given as approximately 100 µg/L in urine [1] (see also Table 1). It should be noted that cotinine determination enables monitoring of only a relatively short period (3–4 days) of previous exposure to tobacco smoke. Measurement of cotinine does not detect smoking that took place before this exposure period.

Table 1. Mean value ranges for cotinine concentrations in the urine of non-smokers, passive smokers and smokers [4, 5]

	Non-smokers	Passive smokers	Smokers	Cut off*
Cotinine in urine [µg/L]	1–10	8–25	1300–1700	60–120

* The cut-off range serves to differentiate between smokers and non-smokers

General characteristics and fundamentals on the toxicology of nicotine and cotinine have already been discussed in this series. Therefore the reader should refer to the 14th issue of "Analysen in biologischem Material" or the 7th volume of "Analysis of Hazardous Substances in Biological Materials" [2, 3].

Author: *M. Müller*
Examiners: *R. Heinrich-Ramm, H.-W. Hoppe*

Cotinine

Application   Determination in urine

Analytical principle   Capillary gas chromatography/
                        mass spectrometric detection (MS)

Completed in   May 2001

Contents

1 General principles
2 Equipment, chemicals and solutions
2.1 Equipment
2.2 Chemicals
2.3 Solutions
2.4 Calibration standards
3 Specimen collection and sample preparation
3.1 Sample preparation
4 Operational parameters
4.1 Operational parameters for gas chromatography and mass spectrometry
5 Analytical determination
6 Calibration
7 Calculation of the analytical result
8 Standardisation and quality control
9 Reliability of the method
9.1 Precision
9.2 Accuracy
9.3 Detection limits
9.4 Sources of error
10 Discussion of the method
11 References

1 General principles

5 M sodium hydroxide and then deuterated cotinine (cotinine-d_3) as an internal standard are added to 2 mL of urine, and extraction is carried out using dichloromethane. The extract is dried under a stream of nitrogen, dissolved in toluene and after capillary gas chromatographic separation, the analyte is quantified by means of mass se-

lective detection in the selective ion monitoring (SIM) mode. Calibration is performed using calibration standards which are prepared in pooled urine and are treated in the same manner as the samples to be analysed.

2 Equipment, chemicals and solutions

2.1 Equipment

Capillary gas chromatograph with split/splitless injector, mass selective detector (MSD) and data processing system

Capillary gas chromatographic column:
Length: 30 m, inner diameter: 0.25 mm; stationary phase: 5% phenylmethylpolysiloxane; film thickness: 0.25 µm (e.g. from Hewlett-Packard)

10 µL Syringe for gas chromatography, but the use of an autosampler is preferable

Glass centrifuge tubes 12 mL (e.g. from Schott) with polyethylene stoppers

Test-tube shaker (Vortex mixer)

Microlitre pipettes, adjustable between 10 and 100 µL, and between 100 and 1000 µL (e.g. from Eppendorf)

Finn pipette 1–5 mL

Laboratory centrifuge

2 mL Autosampler vials

10, 100 and 1000 mL Volumetric flasks

250 mL Glass beaker

Polyethylene sample vessels (2 mL) (e.g. from Eppendorf)

Sterile polyethylene vessels for collection of urine

Polyethylene tubes (10 mL) for portioning

Magnetic stirrer

2.2 Chemicals

Cotinine 98% p.a. (e.g. from Aldrich)

Cotinine-d_3 98 atom % D (e.g. from Sigma)

Dichloromethane p.a. (e.g. from Baker)

Toluene p.a. (e.g. from Fluka)

Sodium sulphate anhydrous p.a. (e.g. from Fluka)

Sodium hydroxide p.a. (e.g. from Fluka)

Concentrated hydrochloric acid p.a. (e.g. from Merck)

Bidistilled water

Helium 4.6 (e.g. from Linde)

Nitrogen 4.0 (e.g. from Linde)

2.3 Solutions

5 M NaOH:
Approximately 150 mL bidistilled water is placed in a 250 mL glass beaker. After the addition of 40 g sodium hydroxide, the solution is mixed with a magnetic stirrer until it is clear, and then the solution is filled to a total volume of 200 mL with bidistilled water.

0.1 M Hydrochloric acid:
10 mL of 37% HCl are placed into a 1000 mL volumetric flask and the flask is filled to its nominal volume with bidistilled water.

Solution of the internal standard

Stock solution:
Approximately 10 mg cotinine-d_3 are weighed exactly into a 100 mL volumetric flask. The flask is subsequently filled to its nominal volume with 0.1 M hydrochloric acid (100 mg/L).

Working solution:
100 µL of the stock solution of the internal standard are pipetted into a 10 mL volumetric flask. The flask is subsequently filled to its nominal volume with 0.1 M hydrochloric acid (1 mg/L).

2.4 Calibration standards

The calibration standard solutions are prepared in pooled urine from non-smoking test persons. For the purpose of preparing pooled urine spontaneous urine samples are collected from the test persons in a suitable vessel, thoroughly mixed and stored at −18 °C until the standards and the control material are prepared. If necessary, the cotinine content of the individual urine samples used to prepare the pooled urine is checked.
The working solution and the calibration standards must be freshly prepared before each analytical series.

Stock solution:
Approximately 10 mg cotinine are weighed exactly into a 100 mL volumetric flask. The flask is subsequently filled to its nominal volume with 0.1 M hydrochloric acid (100 mg/L).

Working solution:
100 µL of the cotinine stock solution are pipetted into a 10 mL volumetric flask. The flask is subsequently filled to its nominal volume with 0.1 M hydrochloric acid (1 mg/L).

Calibration standards:
Calibration standards in pooled urine are prepared from the cotinine working solution and the working solution of the internal standard in accordance with the following pipetting scheme (Table 2).

Table 2. Pipetting scheme for the preparation of the calibration standards

Volume of pooled urine [mL]	Volume of the working solution of the internal standard [µL]	Volume of the cotinine working solution [µL]	Volume of 0.1 M hydrochloric acid [µL]	Concentration of the calibration standard [µg/L]
2	100	–	1400	0
2	100	10	1390	5
2	100	50	1350	25
2	100	100	1300	50
2	100	200	1200	100
2	100	500	900	250

3 Specimen collection and sample preparation

Spontaneous urine samples are collected in sealable polyethylene bottles and stored in the deep-freezer at approx. −18 °C until sample processing for cotinine determination is carried out. The urine can be stored for at least six months under these conditions.

3.1 Sample preparation

Before analysis, the samples are thawed and thoroughly mixed. 100 µL of the working solution of the internal standard and 1400 µL 0.1 M hydrochloric acid are added to 2 mL urine in a glass centrifuge tube and shaken on a Vortex mixer. 2 mL of 5 M sodium hydroxide are added using a pipette and the sample is mixed again (Vortex). Then 5 mL dichloromethane are added to the sample, and the analyte is extracted by shaking for 30 seconds on the Vortex mixer. The phases of the sample contained in the tube sealed with a polyethylene stopper are separated by centrifugation at room temperature for 5 minutes at 1400 g. After carefully withdrawing the upper aqueous phase using a Finn pipette, this phase is discarded. The organic phase is dried by adding 3 g of anhydrous sodium sulphate and by shaking the sample briefly on the Vortex mixer. The sealed sample is subsequently centrifuged anew at room temperature for 5 minutes at 1400 g. The supernatant is transferred to a fresh centrifuge tube and the contents are evaporated carefully to dryness in a stream of nitrogen. Finally the residue is dissolved in 1 mL toluene and transferred into a 2 mL autosampler vial.

4 Operational parameters

4.1 Operational parameters for gas chromatography and mass spectrometry

Capillary column:	Material:	Fused silica
	Stationary phase:	DB-5
	Length:	30 m
	Inner diameter:	0.25 mm
	Film thickness:	0.25 µm
Detector:	Mass selective detector (MSD)	
Temperatures:	Column:	Initial temperature 95 °C for 3 minutes isothermal, then increase at a rate of 5 °C/min to 110 °C, then at a rate of 30 °C/min to 280 °C, then 5 min at the final temperature
	Injector:	250 °C
	Transfer line:	250 °C
Carrier gas:	Helium 4.6 with a constant flow of 1.2 mL per minute	
Split:	Splitless, split on after 30 s	
Sample volume:	2 µL	
Ionisation type:	Electron impact ionisation (EI)	
Ionisation energy:	70 eV	
Dwell time:	50 ms	
Electron multiplier:	1400 V + 600 V rel.	

All other parameters must be optimised in accordance with the manufacturer's instructions.

5 Analytical determination

In each case 2 µL of the toluene solution are injected into the gas chromatograph for the analytical determination of the urine samples processed as described in Section 3.1.

The temporal profiles of the ion traces shown in Table 3 are recorded in the SIM mode.

Table 3. Retention times and masses

Compound	Retention time [min]	Masses
Cotinine	10.2	176
		98*
Cotinine-d_3 (IS)	10.2	179
		101*

The masses marked with * are used for quantitative evaluation.

The retention times shown in Table 3 serve only as a guide. Users of the method must satisfy themselves of the separation power of the capillary column used and the resulting retention behaviour of the substances. Figure 1 shows an example of a chromatogram of the processed urine from a light smoker (5 cigarettes per day).

If the measured values are above the linear range of the calibration graphs (>250 µg/L), the urine samples are diluted with water in the ratio of 1:10, processed and injected anew.

Two quality control samples are analysed with each analytical series.

6 Calibration

The calibration standards are processed in the same manner as the urine samples (Section 3.1) and analysed by means of gas chromatography/mass spectrometry as described in Sections 4 and 5. Linear calibration graphs are obtained by plotting the quotients of the peak areas of cotinine and that of the internal standard as a function of the concentrations used. It is unnecessary to plot a complete calibration curve for every analytical series. It is sufficient to analyse one calibration standard for every analytical series. The ratio of the results obtained for these standards and the result for the equivalent standard in the complete calibration graph is calculated. Using this quotient, each result read off the calibration graph is corrected.

New linear calibration graphs should be plotted if the precision control results indicate systematic deviations.

The calibration graph is linear between the detection limit and 250 µg per litre urine.

7 Calculation of the analytical result

Quotients are calculated by dividing the peak areas of the analyte by that of the internal standard. These quotients are used to read off the pertinent concentration of cotinine in µg per litre from the relevant calibration graph. If the pooled urine used to prepare the calibration standards exhibits a background signal, the resulting calibration graph must be shifted in parallel so that it passes through the zero point of the coordinates. (The concentrations of the background exposure can be read off from the point where the graph intercepts the axis before parallel shifting in each case.)

8 Standardisation and quality control

The guidelines given by the Bundesärztekammer (German Medical Association) [6, 7] and the special preliminary remarks to this series are to be followed to ensure the quality of the analytical results. In order to determine the precision of the method a urine sample containing a constant concentration of cotinine is analysed. As material for quality control is not commercially available, it must be prepared in the laboratory. It is advisable to use the urine or pooled urine of passive smokers for this purpose. The concentration of this control material should lie in the range of the cut-off value of 100 µg/L. A six-month supply of this control material is prepared, divided into aliquots in 10 mL polyethylene tubes which are stored in the deep-freezer. The theoretical value and the tolerance range for this quality control material are determined in a preliminary period (one analysis of the control material on each of 20 different days) [8, 9].

External quality assurance to check the reliability of the method can be achieved by participation in round-robin experiments. The Deutsche Gesellschaft für Arbeits- und Umweltmedizin (German Association for Occupational and Environmental Medicine) offers cotinine as a parameter for toxicological occupational and environmental analyses in their round-robin programme [10].

9 Reliability of the method

9.1 Precision

Pooled urine spiked with 10 µg/L and 100 µg/L cotinine were processed and analysed to check the precision. Seventeen replicate determinations of this urine sample yielded the precision in the series shown in Table 4.

Table 4. Precision for the determination of cotinine

	n	Concentration [µg/L]	Standard deviation (rel.) [%]	Prognostic range [%]
In the series	17	10	6.1	12.9
	17	100	1.3	2.7
From day to day	15	10	6.4	13.6
	15	100	2.8	5.9

In addition, the precision from day to day was determined. The same material was used as for the determination of the precision in the series. This urine was processed and analysed on each of 15 different days. The precision results are also shown in Table 4.

9.2 Accuracy

The loss due to processing was determined to check the accuracy of the method. For this purpose reference standards prepared in water and urine were processed and analysed. Toluene standards were simultaneously prepared with the same cotinine and cotinine-d_3 concentrations as the reference standards in their respective matrix. These toluene standards were injected into the GC and analysed without further treatment. Mean absolute recovery rates of 98.3% (water) and 95.0% (pooled urine) were obtained by comparison of the toluene calibration graphs with the calibration graphs for water and pooled urine. This means that 1.7% cotinine (and cotinine-d_3) are lost during the processing and analysis of aqueous cotinine samples. In the case of urine samples, the losses due to processing are about 5%. As the analyses are carried out using deuterated cotinine as an internal standard, the relative recovery is generally about 100%.

9.3 Detection limits

Under the conditions given here the detection limit, calculated as three times the signal/noise ratio of the analytical background noise in the temporal vicinity of the analyte signal, is approximately 1 µg/L.

9.4 Sources of error

Interference due to matrix effects was not observed for the ion traces *m/z* 98, 101, 176 and 179 under the conditions given here. The deuterated internal standard cotinine-d_3 must be checked to ensure it does not contain impurities of non-deuterated cotinine.

As already ascertained in the alternative method published as part of this series, a liner for the GC injector which has not been properly deactivated or has been contaminated can greatly reduce the sensitivity of the method and can lead to chromatographic problems. Interference to the analysis due to peaks with retention times similar to those of the analyte and internal standard was not observed during the analysis of more than 550 individual urine samples [11].

10 Discussion of the method

The procedure presented here is a further advance on the method devised by Skarping and co-workers in 1988 [12]. It permits the sensitive, reliable and, due to the use of mass spectrometry, extremely specific analysis of cotinine in the concentration range between 1 and 250 µg/L. The procedure enables the determination of cotinine in the urine of non-smokers, passive smokers and persons with a low consumption of tobacco (see also Figure 1).
The urine samples need only be adjusted to the alkaline range, subjected to liquid/liquid extraction and finally analysed. No derivatisation of the analyte is necessary.
The analytical reliability criteria of the method were checked using control material prepared in the laboratory. For this purpose pooled urine of non-smokers was spiked with two different concentrations of cotinine, processed and analysed several times. The precision achieved is excellent. Among other factors this is due to the use of deuterated cotinine as an internal standard. Moreover, the losses due to processing of only 5% in the case of urine and 1.7% for aqueous samples confirm the accuracy of the analytical results. Therefore matrix effects are not to be expected in the analysis of urine samples containing creatinine concentrations in the range between 0.5 and 2.5 g/L. On account of the very slight losses due to processing and the almost 100% relative recovery direct calibration can be carried out without problems instead of a matrix-based calibration.
The method presented here was developed and optimised in order to enable quantification of cotinine in urine without interference in the concentration range between 1 and 250 µg/L. Neither the author nor the examiners observed any interfering peaks, which occasionally occur when a non-specific NPD detector is used in this concentration range [2, 3]. As the cotinine concentration in the urine of smokers is more than 1000 µg/L as a rule, the urine of smokers must be diluted before analysis. For this reason it is advisable to clarify the smoking status of the test person by anamnesis prior to analysis. The linearity range was thoroughly checked by the examiners of the method. Linearity is given up to approx. 1500 µg/L. If calibration is carried out accordingly, many urine samples of smokers can be quantified without prior dilution. The use of mass spectrometry has the advantage of high specificity. This permits a short analysis time.
In a recent study the cotinine values of 569 employees in the construction industry were investigated using this method. There was a good correlation between the smoking habits obtained by anamnesis and the cotinine levels found in the urine of the

test subjects. The method described here has proved successful on account of its practicability for routine investigations and its robustness [11].

Instruments used:
Gas chromatograph 6890 II with mass selective detector 5973, autosampler 6890 and data system from Hewlett-Packard.

11 References

[1] *V. Haufroid* and *D. Lison:* Urinary cotinine as a tobacco-smoke exposure index: A minireview. Int. Arch. Occup. Environ. Health 71, 162–168 (1998)
[2] *G. Scherer* and *I. Meger-Kossien:* Cotinine in urine, plasma or serum. In: *J. Angerer* and *K.H. Schaller (eds.)* Analysis of Hazardous Substances in Biological Materials, Vol. 7. Wiley-VCH, 171–189 (2001)
[3] *G. Scherer* and *I. Meger-Kossien:* Cotinin in Urine, Plasma oder Serum. In: *J. Angerer* and *K.H. Schaller (eds.)* Analysen in Biologischem Material. Loose-leaf collection, 14th Issue. Wiley-VCH (2000)
[4] *N.L. Benowitz:* Cotinine as a biomarker of environmental tobacco smoke exposure. Epidemiol. Rev. 18, 188–204 (1996)
[5] *G. Scherer* and *E. Richter:* Biomonitoring exposure to environmental tobacco smoke (ETS): A critical reappraisal. Hum. Exp. Toxicol. 16, 449–459 (1997)
[6] *Bundesärztekammer:* Qualitätssicherung der quantitativen Bestimmungen im Laboratorium. Neue Richtlinien der Bundesärztekammer. Dt. Ärztebl. 85, A699–A712 (1988)
[7] *Bundesärztekammer:* Ergänzung der „Richtlinien der Bundesärztekammer zur Qualitätssicherung in medizinischen Laboratorien". Dt. Ärztebl. 91, C159–C161 (1994)
[8] *G. Lehnert, J. Angerer* and *K.H. Schaller:* Statusbericht über die externe Qualitätssicherung arbeits- und umweltmedizinisch-toxikologischer Analysen in biologischen Materialien. Arbeitsmed. Sozialmed. Umweltmed. 33(1), 21–26 (1998)
[9] *J. Angerer* and *G. Lehnert:* Anforderungen an arbeitsmedizinisch-toxikologische Analysen – Stand der Technik. Dt. Ärztebl. 37, C1753–C1760 (1997)
[10] *Ringversuch Nr. 28*: Qualitätsmanagement in der Arbeits- und Umweltmedizin, Projektgruppe Qualitätssicherung, Organisation: Institut für Arbeits-, Sozial- und Umweltmedizin der Universität Erlangen-Nürnberg (2001)
[11] *M. Müller, P. Ruhnau, C. Caumanns, R. Böhm* and *E. Hallier:* Cotininbestimmung aus dem Humanurin zur Objektivierung des Confounders Rauchen in potentiell belasteten Kollektiven mit GC/MS. Verh. Dt. Ges. Arbeitsmed. 39, 651–653 (1999)
[12] *G. Skarping, S. Willers* and *M. Dalene:* Determination of cotinine in urine using glass capillary gas chromatography and selective detection, with special reference to the biological monitoring of passive smoking. J. Chromatogr. 454, 293–301 (1988)

Author: *M. Müller*
Examiners: *R. Heinrich-Ramm, H.-W. Hoppe*

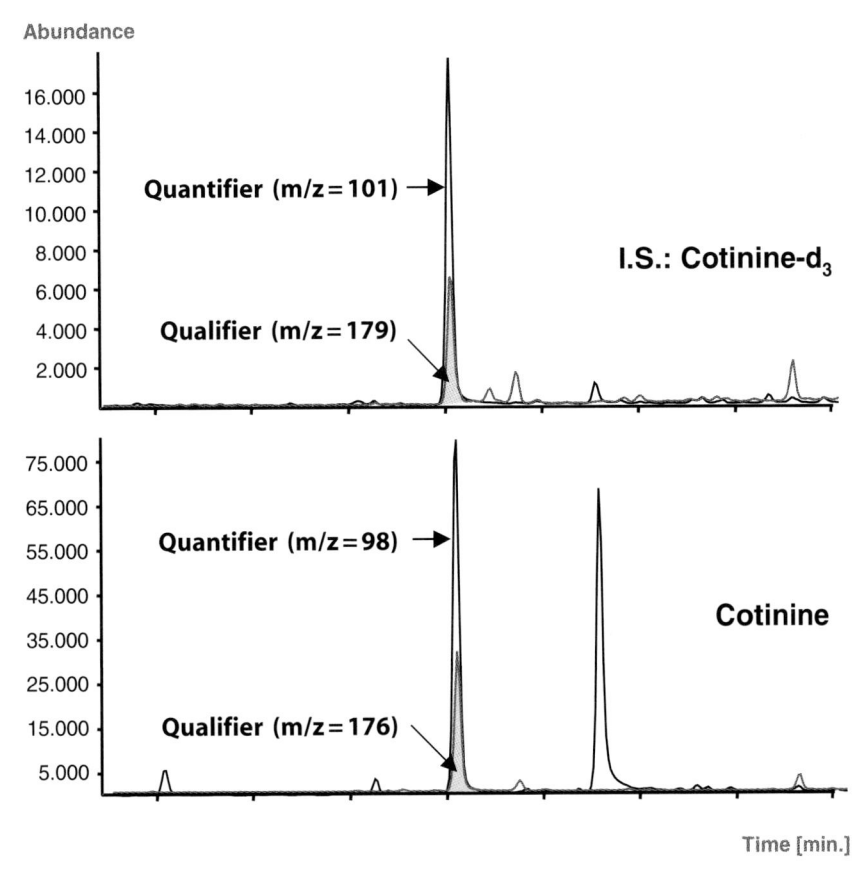

Fig. 1. Example of a chromatogram of a light smoker (5 cigarettes per day) with a cotinine concentration of 227 µg/L urine

N,N-Dimethylacetamide (DMA) and N-methylacetamide (NMA)

Application   Determination in urine

Analytical principle   Capillary gas chromatography/nitrogen/phosphorus-specific detector (NPD)

Completed in   March 2002

Summary

This method permits the simultaneous determination of N,N-dimethylacetamide (DMA) and its metabolite N-methylacetamide (NMA) in urine after occupational exposure to DMA.

DMA is metabolised and excreted in the human organism in the same manner as N,N-dimethylformamide (DMF) [1]. In addition to the non-metabolised DMA, its metabolite, N-hydroxymethyl-N-methylacetamide (HMMA) is of importance for biological monitoring in urine. It can be assumed that HMMA is thermally decomposed to formaldehyde and N-methylacetamide in the injector of the gas chromatograph during chromatographic determination, as is the case for the HMMA homologue N-hydroxymethyl-N-methylformamide [2, 3]. After capillary gas chromatographic separation, the analysis is carried out by means of a nitrogen/phosphorus-specific detector (NPD). N,N-Dimethylpropanamide (DMPA) serves as an internal standard. Calibration is carried out using calibration standards prepared in urine and treated in the same manner as the samples to be tested. The quality criteria were determined using urine samples of test subjects who had been exposed to DMA.

N,N-Dimethylacetamide (DMA)

Within-series imprecision: Standard deviation (rel.) s_w = 6.7% or 5.1%
                            Prognostic range u = 14.9% or 11.4%
                            at a concentration of 21 µg or 410 µg DMA per litre urine
                            and where n = 10 determinations

Between-day imprecision: Standard deviation (rel.) s_w = 7.2% or 6.0%
                            Prognostic range u = 16.0% or 13.4%
                            at a concentration of 21 µg or 410 µg DMA per litre urine
                            and where n = 10 determinations

N,N-Dimethylacetamide (DMA) and N-methylacetamide (NMA)

Accuracy: Recovery rate $r = 86.4–109.1\%$ at 0.1 mg/L

Detection limit: 10 µg DMA per litre urine

N-Methylacetamide (NMA)

Within-series imprecision: Standard deviation (rel.) $s_w = 2.3\%$ or 2.9%
Prognostic range $u = 4.7\%$ or 5.9%
at a concentration of 280 µg or 8300 µg NMA per litre urine
and where $n = 10$ determinations

Between-day imprecision: Standard deviation (rel.) $s_w = 4.0\%$ or 3.8%
Prognostic range $u = 8.2\%$ or 7.8%
at a concentration of 280 µg or 8300 µg NMA per litre urine
and where $n = 10$ determinations

Accuracy: Recovery rate $r = 91.7–113.6\%$ at 1.0 mg/L

Detection limit: 10 µg NMA per litre urine

N,N-Dimethylacetamide (DMA)

Dimethylacetamide is a colourless liquid (MW: 87.12; density: 0.943 kg/L) which is miscible with water and most polar organic solvents. Its melting point is $-20\,°C$, DMA boils at $166\,°C$.

DMA is used as a polar solvent for polyacrylonitrile and other polymers. Furthermore, it serves as a solvent for gases, as a stripping agent and extraction agent, as a catalyst and crystallisation agent, and at an ultrapure grade of purity as a solvent for spectroscopy.

DMA is described as a solvent with low acute toxicity and a slight potential for irritation [4, 5]. Target organs for systemic effects are mainly the liver and the testes.

DMA is classified in the list of MAK values as a substance with a risk of absorption, with the onset of symptoms within 2 hours and a half-life of <2 hours. The currently valid MAK value in Germany is 10 ppm or 36 mg/m^3. In addition to adsorption by inhalation, intake through the skin can make a decisive contribution to the inner exposure of workers, or it can even be the most important route of intake. Therefore the substance is marked with an "H" (for absorption through the skin) in the list of MAK values. In the justifications for MAK values the question of embryotoxic and teratogenic effects has been comprehensively addressed from the point of view of occupational medicine and toxicology [6]. According to the conclusions reached there, the risk of damage to embryos and foetuses from DMA needs not be feared provided compliance with the MAK and BAT values is ensured at the workplace.

The metabolism of DMA is analogous to that of DMF [7]. Figure 1 shows the metabolism of DMA [1]. N-methylacetamide (NMA) is excreted as the main metabolite in addition to the non-metabolised substance (DMA). The concentration of this metabolite results partly from the thermal decomposition of N-methyl-N-hydroxymethylacetamide in the GC injector and partly from the metabolisation of DMA. Both

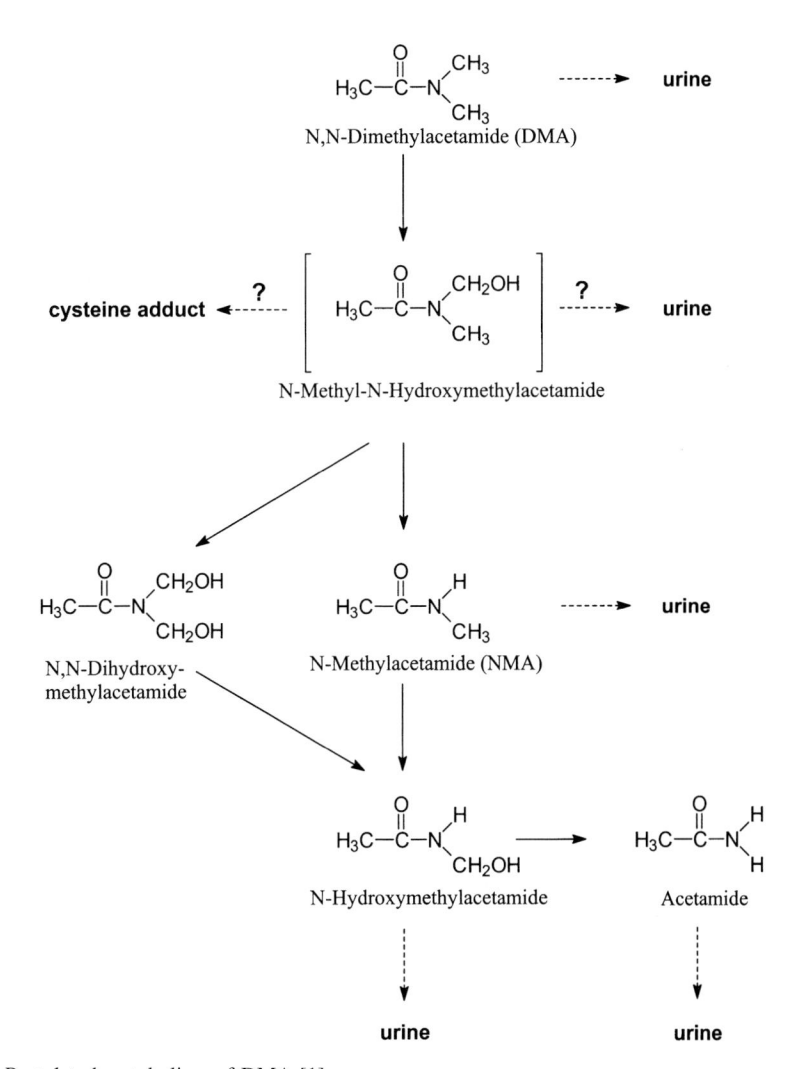

Fig. 1. Postulated metabolism of DMA [1]

Table 1. Data from field and laboratory studies on the external and internal exposure levels to N,N-dimethylacetamide

Type of study	External exposure to DMA	NMA in urine	Remarks	Reference
Laboratory study (n = 2)	10 ppm; 6 h With and without a respiratory protection mask	25 mg/L or 30 mg/g creatinine	Maximum excretion 5–10 h after end of exposure; 70% resulting from inhalation, 30% from cutaneous absorption	Maxfield et al. 1975 [7]
Field study (n = 8)	5 days; 8 h exposure to 14 ppm (mean value)	48 mg/L or 57 mg/g creatinine (mean value)	Steady state after 2–3 days; dermal contact was not prevented; no correlation between NMA in the urine after a shift and DMA in the air; $t_{½}$ = 16 h	Borm et al. 1987 [15]
Field study (n = 5)	4 weeks; mean value: 1.2 ppm	13 mg/L (mean value); range: 1 to 42 mg/L	No marked accumulation over the working week; there was direct contact with the skin; no extrapolation to higher air concentrations	Kennedy et al. 1989 [16]
Field study (n = 93)	2 days of exposure (12 h in each case) after 3 days without exposure: 1.45 ± 2.25 ppm (geometric mean)	88 mg/g creatinine	Increase of NMA between the 1st and 2nd day (accumulation)	Spies et al. 1995 [17]
Laboratory study (n = 10)	8 h at 10 ppm; with and without respiratory protection mask; physical exertion: bicycle ergometer: 10 min/h at 75 Watt	21.4 ± 10.2 mg/g creatinine 95th percentile: 31.9 mg/g creatinine	Half-life of the NMA elimination: $t_{½}$ = 9.2 ± 3.6 h; absorption through the skin amounts to approx. 13 to 24% of the total intake.	Knecht et al. 1997 [18]

DMA and NMA are suitable for biological monitoring. Results from field and laboratory studies on the relationship between an occupational outer and inner exposure to DMA are described in the literature (Table 1). No data on possible general background levels have been published as yet.

A BAT value has been assigned to DMA. It was stipulated as 30 mg NMA/g creatinine [8]. The American BEI value for NMA is given as 30 mg/g creatinine at the end of the last shift in a working week [9].

Author: *U. Knecht*
Examiner: *G. Müller*

N,N-Dimethylacetamide (DMA) and N-methylacetamide (NMA)

Application   Determination in urine

Analytical principle Capillary gas chromatography/nitrogen/phosphorus-specific detector (NPD)

Completed in   March 2002

Contents

1   General principles
2   Equipment, chemicals and solutions
2.1 Equipment
2.2 Chemicals
2.3 Solutions
2.4 Calibration standards
3   Specimen collection and sample preparation
3.1 Sample preparation
4   Operational parameters for gas chromatography
5   Analytical determination
6   Calibration
7   Calculation of the analytical result
8   Standardisation and quality control
9   Reliability of the method
9.1 Precision
9.2 Accuracy
9.3 Detection limits
9.4 Sources of error
10 Discussion of the method
11 References

1 General principles

The DMA metabolite N-hydroxymethyl-N-methylacetamide (HMMA) is determined in urine in addition to the non-metabolised DMA. It can be assumed that HMMA is thermally decomposed to formaldehyde and N-methylacetamide (NMA) in the injector of the gas chromatograph during chromatographic determination, as is the case for the HMMA homologue N-hydroxymethyl-N-methylformamide [2, 3]. After capillary gas chromatographic separation, the analysis is carried out by means of a nitrogen/phosphorus detector (NPD). N,N-Dimethylpropanamide (DMPA) serves as an internal standard. Calibration is carried out using calibration standards prepared in urine and treated in the same manner as the samples to be tested. The quality criteria were determined using urine samples of test subjects who had been exposed to DMA.

2 Equipment, chemicals and solutions

2.1 Equipment

Capillary gas chromatograph with nitrogen/phosphorus detector (NPD) with manual or automatic sample injection as well as an integrator or a PC system for data evaluation.

Separation column: Stabilwax (e.g. from Restek; equivalent to a Carbowax column PEG); length: 60 m, inner diameter: 0.53 mm, film thickness: 1.5 µm

5 µL Syringe for gas chromatography

Laboratory shaker (e.g. IKA KS 500)

Centrifuge (e.g. from Heraeus)

Volumetric flasks: 100 mL, 50 mL and 10 mL

2 mL Pipettes

1 mL Measurement pipettes

Microlitre tips: 500 µL and 50 µL

10 mL Centrifuge vials

10 mL Polyethylene tubes

2.2 Chemicals

N,N-Dimethylacetamide p.a. (e.g. from Fluka)

N-Monomethylacetamide p.a. (e.g. from Aldrich)

N,N-Dimethylpropanamide p.a. (e.g. from Aldrich)

1-Butanol, p.a. (e.g. from Merck)

Deionised water (e.g. produced by means of Millipore® technology)

2.3 Solutions

Stock solution of the internal standard (IS):
Approximately 100 mg DMPA are weighed exactly into a 10 mL volumetric flask. The flask is subsequently filled to its nominal volume with water (10 g/L).

Spiking solution for the internal standard:
100 µL of the stock solution of the internal standard are pipetted into a 100 mL volumetric flask. The flask is subsequently filled to its nominal volume with water (10 mg/L).

2.4 Calibration standards

Stock solution of DMA:
Approximately 100 mg DMA are weighed exactly into a 100 mL volumetric flask. The flask is subsequently filled to its nominal volume with water (1000 mg/L).

Stock solution of NMA:
Approximately 100 mg NMA are weighed exactly into a 100 mL volumetric flask. The flask is subsequently filled to its nominal volume with water (1000 mg/L).

Calibration standards (100–10,000 µg/L)

Calibration standards in pooled urine are prepared from the stock solutions in accordance with the following pipetting scheme. For this purpose the volumes of the stock solutions shown in Table 2 are each filled into a 50 mL volumetric flask, and then the flask is filled to its nominal volume with pooled urine. This material is divided into 3 mL aliquots which are pipetted into 10 mL sealable polyethylene tubes and stored at approx. –18 °C. Calibration standards thus stored are stable for at least 12 months.

Table 2. Pipetting scheme for the preparation of calibration standards in pooled urine in 50 mL volumetric flasks

Volume of the DMA stock solution [µL]	Volume of the NMA stock solution [µL]	Concentration of the calibration standard [µg/L]
–	–	0
5	5	100
25	25	500
50	50	1000
125	125	2500
250	250	5000
375	375	7500
500	500	10,000

3 Specimen collection and sample preparation

The specimens should be collected at the end of a working shift. The urine samples are collected in plastic bottles. If the samples cannot be processed immediately, the urine must be stored in the deep-freezer at approx. −18 °C until it can be processed. Samples thus stored are stable for at least 12 months.

3.1 Sample preparation

Before analysis, the samples are thawed (if necessary) and thoroughly mixed. 2 mL urine are pipetted into a 10 mL centrifuge vial and 200 µL of the spiking solution of the internal standard are added. 2 mL 1-butanol are added using a pipette, and the sample is shaken vigorously for 10 minutes on a sample shaker. Then the sample is centrifuged for 5 minutes at 3500 g to separate the phases.
Approximately 500 µL of the upper 1-butanol phase are withdrawn using a pipette and transferred to an autosampler vial for the subsequent analysis.

4 Operational parameters for gas chromatography

Capillary column:	Material:	Fused silica
	Stationary phase:	Carbowax
	Length:	60 m
	Inner diameter:	0.53 mm
	Film thickness:	1.5 µm
Detector:	Nitrogen/phosphorus-specific detector (NPD)	
Reaction gas:	Hydrogen 5.0 at a flow rate of 0.6 mL/min	
Temperatures:	Column:	180 °C isothermal
	Injector:	240 °C
	Detector:	260 °C
Carrier gas:	Helium 5.0 at a flow rate of 3.7 mL/min	
Split:	1:10	
Injection volume:	1 µL	

All other parameters must be optimised in accordance with the manufacturer's instructions.
The retention times given below were found for the analytes and the internal standard under the chromatographic conditions described here. They serve only as a guide. Users of the method must satisfy themselves of the separation power of the GC column they use and of the resulting retention behaviour of the substances.

Retention times:
N,N-Dimethylacetamide: 13.86 minutes
N-Methylacetamide: 17.33 minutes
N,N-Dimethylpropanamide (IS):   14.39 minutes

Figure 2 shows the chromatogram of a processed urine sample from a test person who had been exposed to DMA.

5 Analytical determination

In order to carry out the gas chromatographic analysis, 1 µL of each of the prepared urine samples is injected into the GC. If the measured values are outside the range of the calibration graph, the samples are appropriately diluted and processed anew.
A quality control sample is analysed with each analytical series.

6 Calibration

The calibration standards are processed in the same manner as the urine samples (Section 3.1) and analysed by means of capillary gas chromatography with nitrogen-specific detection as described in Sections 4 and 5. In each case 1 µL of the processed calibration standards in urine is injected into the GC.
Linear calibration graphs are obtained by plotting the quotients of the peak areas of DMA or NMA and that of the internal standard as a function of the concentrations used.
Calibration must be carried out for each analytical series on account of the susceptibility of the NPD to interference.
Urine samples with a DMA or NMA concentration of between 10 µg/L and 10,000 µg/L can be measured under the conditions described here and using this detector.
Calibration graphs of DMA and NMA are shown in Figure 3 and Figure 4.

7 Calculation of the analytical result

Quotients are calculated by dividing the peak areas of the analyte by that of the internal standard. These quotients are used to read off the concentration of DMA or NMA in µg per litre urine from the relevant calibration graph. If the pooled urine used to prepare the calibration standards exhibits a background signal, the resulting calibration graph must be shifted in parallel so that it passes through the zero point of the coordinates. (The concentrations of the background exposure can be read off from the point where the graph intercepts the axis before parallel shifting in each case.) Any reagent blank values must be subtracted from the analytical results for the real samples.

8 Standardisation and quality control

Quality control of the analytical results is carried out as stipulated in the guidelines of the Bundesärztekammer (German Medical Association) [10, 11] and in the special preliminary remarks to this series. In order to determine the precision of the method a urine sample containing a constant concentration of DMA and NMA is analysed. As material for quality control is not commercially available, it must be prepared in the laboratory. For this purpose, pooled urine is spiked with a defined quantity of DMA and NMA. The concentration of this control material should lie within the relevant concentration range. A six-month supply of the control material is prepared, divided into aliquots in 10 mL polyethylene tubes and stored in the deep-freezer. Control material thus stored is stable for at least 12 months. The theoretical value and the tolerance range for this quality control material are determined in the course of a pre-analytical period (one analysis of the control material on each of 20 different days) [12–14].

9 Reliability of the method

9.1 Precision

The precision in the series was determined by processing and analysing two individual urine samples from test persons who had been exposed to DMA. Ten replicate determinations of the urine samples yielded the precision in the series shown in Table 3.

Table 3. Precision in the series for the determination of DMA and NMA (n = 10)

Substance	Concentration [µg/L]	Standard deviation (rel.) [%]	Prognostic range [%]
DMA	21	6.7	14.9
	410	5.1	11.4
NMA	280	2.3	4.7
	8300	2.9	5.9

In addition, the precision from day to day was determined. The same material was used as for the determination of the precision in the series. These urine samples were processed and analysed on each of 10 different days. The precision from day to day for DMA and NMA is given in Table 4.

Table 4. Precision from day to day for the determination of DMA and NMA (n = 10)

Substance	Concentration [µg/L]	Standard deviation (rel.) [%]	Prognostic range [%]
DMA	21	7.2	16.0
	410	6.0	13.4
NMA	280	4.0	8.2
	8300	3.8	7.8

9.2 Accuracy

Recovery experiments were performed to check the accuracy of the method. A urine sample of known DMA and NMA concentration was processed and analysed 10 times in accordance with Section 3. The mean recovery rates were between 86.4 and 109.1% for a DMA concentration of 0.1 mg/L and between 91.7 and 113.6% for a NMA concentration of 10 mg/L urine.

In addition, the losses due to processing were checked as a measure of the accuracy. For this purpose reference standards prepared in urine were processed and analysed. Standards in 1-butanol containing the same amounts of DMA and NMA as the reference standards in urine were simultaneously prepared. These 1-butanol standards were injected into the GC and analysed without further treatment. Comparison of the peak areas of the 1-butanol standards with the corresponding peak areas of the processed reference standards in urine showed mean absolute recovery rates of 48% for DMA or 51% for NMA at a concentration of 1000 µg/L. This means that on average 52% of DMA and 49% of NMA are lost during processing of the urine samples. Addition of sodium chloride to the urine sample reduces these losses to 29% and 30% respectively.

9.3 Detection limits

The detection limit is approx. 10 µg DMA or NMA per litre urine under the conditions for sample processing and chromatographic determination described here. The detection limit was calculated on the basis of three times the signal-background ratio.

9.4 Sources of error

The sample processing ensures that no interference occurs in chromatographic separation. The chromatogram is notable for its stable baseline. An interfering background was not detected, even in urine samples with a very low concentration of DMA or NMA. An insufficient separation of the peak following the NMA signal (see also Figure 2) must be prevented, otherwise false positive NMA concentrations may result. Furthermore, it is essential to ensure that only gases of the highest purity are used with the NPD. The measurement system of the NPD is very sensitive and requires special maintenance and care. This applies especially to all the parts of the detector that come into contact with the sample material to be determined, e.g. glass parts and seals as well as the electrode in particular. These parts must be heated to glowing in hydrogen 5.0 before the detector is put into operation. Conventional combustible gases are unsuitable for this purpose. There are no restrictions on the use of the detector as long as it is maintained in a stable state. The operator must follow the manufacturer's recommendations. It should not be switched off, even if it is not in use for a longer period of time. The extraction agent, butanol, may cause negative peaks in the chromatogram.

10 Discussion of the method

Determination of the sum of N-methyl-N-hydroxymethylacetamide and NMA as well as DMA in the urine of workers after a working shift proved to be an extremely practicable method of estimating the absorbed DMA. The method presented here permits the reliable, sensitive and accurate determination of the DMA metabolites in the urine of persons who are exposed to DMA at the workplace.

The examiner of the method was able to duplicate the procedure with ease and confirms the reliability criteria given here. A comparison between calibration using aqueous standards and standards prepared in urine showed only slight differences in the gradients of the resulting calibration functions. In view of the losses due to processing and the observed effect after addition of sodium chloride, the internal standard proved to be an eminently suitable choice.

To the best of our current knowledge the method presented here is the most sensitive of the methods available at present. In addition, sample processing can be carried out with a minimum of effort. Therefore an experienced laboratory technician is capable of processing and analysing 40–45 samples a day.

Furthermore, this method is also suitable for simultaneous determination of N,N-dimethylformamide (DMF), which is excreted renally without change after exposure to N,N-dimethylformamide, its metabolite N-methylformamide, as well as DMA and NMA with the same sensitivity, while retaining N,N-dimethylpropanamide as the internal standard. Alternatively, mass spectrometry may be used as the detection technique.

Instruments used:
Sichromat 2 capillary gas chromatograph with nitrogen/phosphorus detector (NPD) from Siemens.

11 References

[1] *R.R. Lauwerys* and *P. Hoet:* Industrial Chemical Exposure: Guidelines for Biological Monitoring. Lewis Publishers, Boca Raton, Florida (1993)
[2] *V. Scailteur* and *R. Lauwerys:* In vivo metabolism of dimethylformamide and relationship to toxicity in male rats. Arch. Toxicol. 56, 87–91 (1984)
[3] *V. Scailteur* and *R. Lauwerys:* In vivo and in vitro oxidative biotransformation of dimethylformamide in rats. Chem. Biol. Interact. 50, 327–337 (1984)
[4] *A. Gescher* and *M.E. Threadgill:* Browning's toxicity and metabolism of industrial solvents. 2nd ed. Elsevier, Amsterdam (1990)
[5] *G. Kennedy:* Biological effects of acetamide, formamide, and their monomethyl and dimethyl derivatives. CRC Crit. Rev. Toxicol. 17, 129–182 (1986)
[6] *H. Greim (ed.):* N,N-Dimethylacetamid. Toxikologisch-arbeitsmedizinische Begründungen von MAK-Werten. Wiley-VCH, Weinheim; 26th issue (1998)
[7] *M. Maxfield, J. Barnes, A. Azar* and *H. Trochimovicz:* Urinary excretion of metabolites following human exposure to DMF and DMAC. J. Occup. Med. 17, 506–511 (1975)
[8] *Deutsche Forschungsgemeinschaft:* MAK- und BAT-Werte-Liste, Report 37. Wiley-VCH, Weinheim (2001)
[9] *ACGIH – American Conference of Governmental Industrial Hygienists:* Threshold limit values for chemical substances and physical agents and biological exposure indices. Kemper Meadow Drive, Cincinnati (1996)
[10] *Bundesärztekammer:* Qualitätssicherung der quantitativen Bestimmungen im Laboratorium. Neue Richtlinien der Bundesärztekammer. Dt. Ärztebl. 85, A699–A712 (1988)
[11] *Bundesärztekammer:* Ergänzung der „Richtlinien der Bundesärztekammer zur Qualitätssicherung in medizinischen Laboratorien". Dt. Ärztebl. 91, C159–C161 (1994)
[12] *G. Lehnert, J. Angerer* and *K.H. Schaller:* Statusbericht über die externe Qualitätssicherung arbeits- und umweltmedizinisch-toxikologischer Analysen in biologischen Materialien. Arbeitsmed. Sozialmed. Umweltmed. 33(1), 21–26 (1998)
[13] *J. Angerer* and *G. Lehnert:* Anforderungen an arbeitsmedizinisch-toxikologische Analysen – Stand der Technik. Dt. Ärztebl. 37, C1753–C1760 (1997)
[14] *J. Angerer, T. Göen* and *G. Lehnert:* Mindestanforderungen an die Qualität von umweltmedizinisch-toxikologischen Analysen. Umweltmed. Forsch. Prax. 3, 307–312 (1998)
[15] *P.J.A. Borm, L. de Jong* and *A. Vliegen:* Environmental and biological monitoring of workers occupationally exposed to dimethylacetamide. J. Occup. Med. 29, 898–903 (1987)
[16] *G.L. Kennedy* and *J.W. Pruett:* Biological monitoring for dimethylacetamide: measurement for 4 consecutive weeks in a workplace. J. Occup. Med. 31, 47–50 (1989)
[17] *G.J. Spies, R.H. Rhyne, K.E. Wetzel, D.T. Ragland, H.G. Turney, T.L. Leet* and *J.L. Oglesby:* Monitoring acrylic fibre workers for liver toxicity and exposure to dimethylacetamide. 1. Assessing exposure to dimethylacetamide by air and biological monitoring. J. Occup. Environ. Med. 37, 1093–1101 (1995)
[18] *U. Knecht:* Standardisierte Untersuchungen zur dermalen Gasphasen-Resorption am Beispiel des Lösungsmittels N,N-Dimethylacetamid. In: *A. Harwerth (ed.):* Verband Deutscher Betriebs- und Werksärzte e.V., Conference Report 1997. Gentner-Verlag, Stuttgart, 153–155 (1997)

Author: *U. Knecht*
Examiner: *G. Müller*

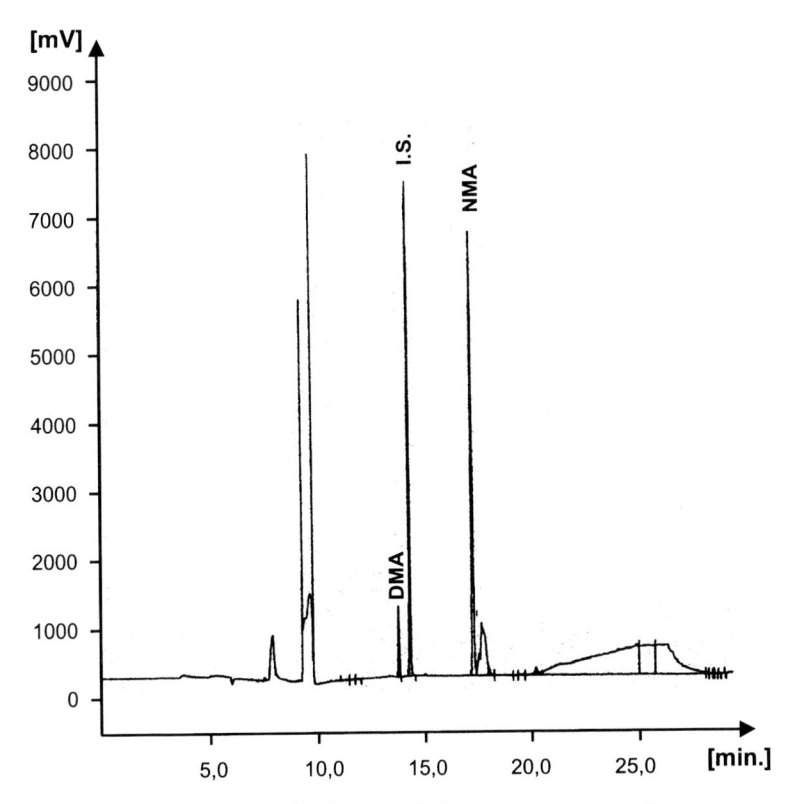

Fig. 2. Chromatogram of a processed urine sample from a test person after exposure to DMA ($c_{DMA} = 1.1$ mg/L, $c_{NMA} = 14.8$ mg/L)

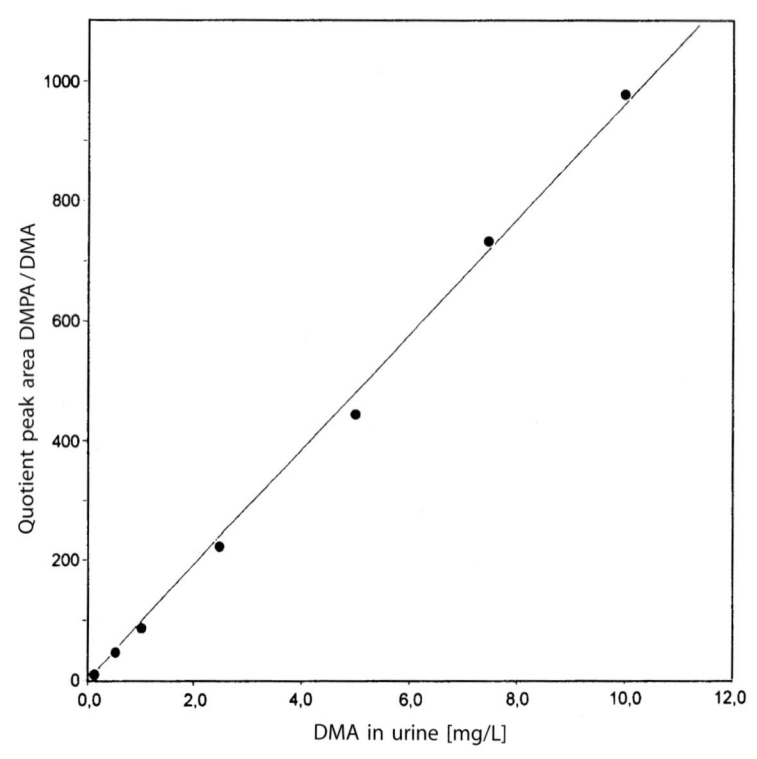

Fig. 3. Calibration function for DMA

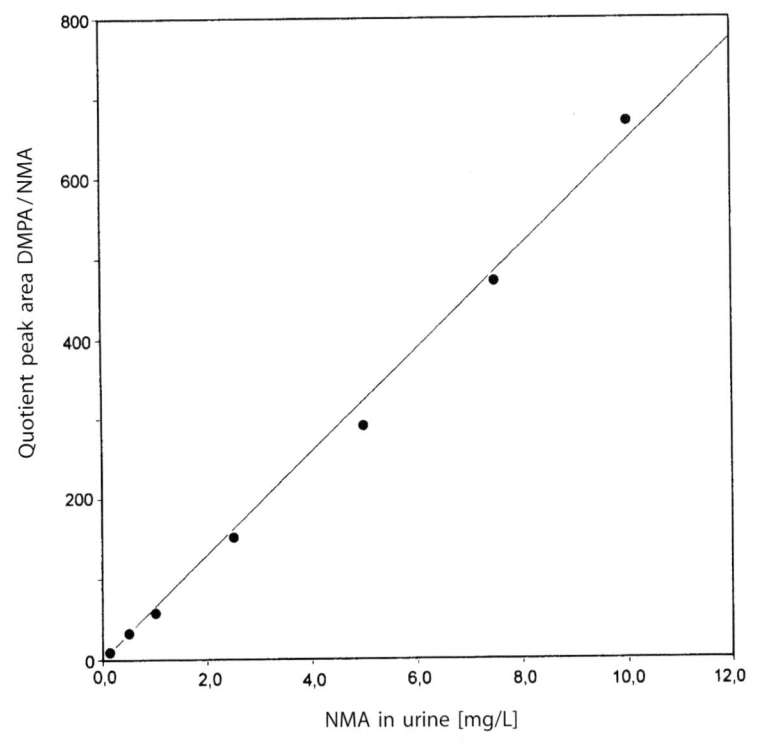

Fig. 4. Calibration function for NMA

Dioxins, furans and WHO PCB in whole blood

Application   Determination in whole blood

Analytical principle   Capillary gas chromatography/
mass spectrometric detection (MS)

Completed in   May 2001

Summary

The method described here permits the sensitive and specific quantitative determination of the tetra-, penta-, hepta-, and octachlorodibenzodioxins and furans with 2,3,7,8-chlorine substitution patterns as well as the coplanar and non-*ortho*-substituted polychlorinated biphenyls (WHO PCB).

The method is based on a specific extraction of the blood fat from 40 g of whole blood, followed by a solid-phase clean-up of the extract on modified silica gels, active carbon and aluminium oxide. The quantitative determination is carried out by means of capillary gas chromatography in combination with high-resolution mass spectrometry in the SIM mode. Calibration standards are prepared in toluene and are measured without further processing. The ^{13}C-labelled compounds corresponding to the analytes serve as internal standards. The toxicity equivalents based on 2,3,7,8-tetrachlorodibenzodioxin are calculated from the measured values for the individual congeners with the help of the WHO calculation model.

2,3,7,8-Tetrachlorodibenzodioxin (2,3,7,8-TCDD)

Within-series imprecision: Standard deviation (rel.) $s_w = 9.5\%$
                                Prognostic range $u = 23.7\%$
                                at a concentration of 2.4 pg 2,3,7,8-TCDD per gram of blood fat
                                and where $n = 6$ determinations

Between-day imprecision: Standard deviation (rel.) $s_w = 22.7\%$
                                Prognostic range $u = 48.1\%$
                                at a concentration of 2.0 pg 2,3,7,8-TCDD per gram of blood fat
                                and where $n = 16$ determinations

Detection limit: 1 pg 2,3,7,8-TCDD per gram of blood fat

1,2,3,6,7,8-Hexachlorodibenzodioxin (1,2,3,6,7,8-HxCDD)

Within-series imprecision: Standard deviation (rel.) $s_w = 3.9\%$
Prognostic range $u = 9.7\%$
at a concentration of 28 pg 1,2,3,6,7,8-HxCDD per gram of blood fat
and where $n = 6$ determinations

Between-day imprecision: Standard deviation (rel.) $s_w = 19.9\%$
Prognostic range $u = 42.2\%$
at a concentration of 19 pg 1,2,3,6,7,8-HxCDD per gram of blood fat
and where $n = 17$ determinations

Detection limit: 1 pg 1,2,3,6,7,8-HxCDD per gram of blood fat

2,3,4,7,8-Pentachlorodibenzofuran (2,3,4,7,8-PeCDF)

Within-series imprecision: Standard deviation (rel.) $s_w = 4.7\%$
Prognostic range $u = 11.7\%$
at a concentration of 13 pg 2,3,4,7,8-PeCDF per gram of blood fat
and where $n = 6$ determinations

Between-day imprecision: Standard deviation (rel.) $s_w = 15.5\%$
Prognostic range $u = 32.8\%$
at a concentration of 12 pg 2,3,4,7,8-PeCDF per gram of blood fat
and where $n = 17$ determinations

Detection limit: 1 pg 2,3,4,7,8-PeCDF per gram of blood fat

Sum of 2,3,7,8-PCDD and PCDF (WHO TEQ)

Within-series imprecision: Standard deviation (rel.) $s_w = 3.5\%$
Prognostic range $u = 8.7\%$
at a concentration of 20 pg per gram of blood fat
and where $n = 6$ determinations

Between-day imprecision: Standard deviation (rel.) $s_w = 13.9\%$
Prognostic range $u = 29.4\%$
at a concentration of 18 pg per gram of blood fat
and where $n = 17$ determinations

3,3',4,4',5-Pentachlorobiphenyl (3,3',4,4',5-PeCB, PCB 126)

Within-series imprecision: Standard deviation (rel.) $s_w = 1.9\%$
Prognostic range $u = 4.7\%$
at a concentration of 71 pg 3,3',4,4',5-PeCB per gram of blood fat
and where $n = 6$ determinations

Between-day imprecision: Standard deviation (rel) $s_w = 25.6\%$
Prognostic range $u = 54.2\%$
at a concentration of 71 pg 3,3',4,4',5-PeCB per gram of blood fat
and where $n = 17$ determinations

Detection limit: 10 pg 3,3',4,4',5-PeCB per gram of blood fat

2,3',4,4',5-Pentachlorobiphenyl (2,3',4,4',5-PeCB, PCB 118)

Within-series imprecision: Standard deviation (rel.) $s_w = 3.2\%$
Prognostic range $u = 8.0\%$
at a concentration of 18,000 pg 2,3',4,4',5-PeCB per gram of blood fat
and where $n = 6$ determinations

Between-day imprecision: Standard deviation (rel.) $s_w = 11.8\%$
Prognostic range $u = 25.0\%$
at a concentration of 13,000 pg 2,3',4,4',5-PeCB per gram of blood fat
and where $n = 17$ determinations

Detection limit: 10 pg 2,3',4,4',5-PeCB per gram of blood fat

Dibenzodioxins and dibenzofurans

Polychlorinated dibenzodioxins (PCDDs) and dibenzofurans (PCDFs) are tricyclic aromatic ethers of extraordinary stability. They are usually grouped together and referred to as "dioxins" in everyday language.

Dibenzodioxins (PCDD)   Dibenzofurans (PCDF)   x,y = 1-4

Of the total of 210 possible congeners and isomers (75 dioxins and 135 furans), the substances with chlorine substitution on at least positions 2, 3, 7 and 8 are of special significance on account of their toxicity. The best-known representative of this group is 2,3,7,8-tetrachlorodibenzo[b,e][1,4]dioxin (2,3,7,8-TCDD) which is also known to the general public as *Seveso Dioxin*.

2,3,7,8-Tetrachlorodibenzodioxin
(2,3,7,8-TCDD)

2,3,7,8-Tetrachlorodibenzofuran
(2,3,7,8-TCDF)

2,3,7,8-TCDD has proved extremely toxic in all the animal species investigated to date; but relatively large differences between the species have been ascertained. When 2,3,7,8-TCDD is administered to rats, they develop the *wasting syndrome*, among other disorders. This is associated with a great reduction in weight due to diminished food intake. The test animals finally die of emaciation in connection with a sharp fall in their blood sugar values.

In humans the most important symptom of acute intoxication or long-term exposure to relatively high doses at the workplace is chloracne.

In vivo and *in vitro* experiments showed that PCDD and PCDF are not genotoxic, but they are carcinogenic. Therefore it is assumed that these substances can to some extent act via very different epigenetic, e.g. tumour-promoting mechanisms [1–4].

In contrast to other chlorinated organic compounds, dioxins were never synthesised for technical use, but are formed as undesired by-products in many thermal and chemical processes. Dioxins and furans were largely introduced into the environment via the air from this primary source. Almost all congeners are detectable in normal environmental samples – water, sediment, soil and air. The source of the dioxins can often be deduced from the relative proportions of the individual congeners in the total dioxin content, which shows a typical pattern.

Almost without exception it is the congeners with 2,3,7,8-substitution that are detected in the human organism, as these dioxins are relatively slowly metabolised and eliminated, thus they are bioaccumulated. If other congeners/isomers are determined in human material, the person must have been exposed very recently – or the sample must have been contaminated on the way to the laboratory.

The main source of human exposure to dioxin is food and in particular fatty food of animal origin [5–7]. As a result of the present daily mean intake of 0.72 (0.18 to 1.7) pg international toxicity equivalents (I-TEQ) per kg body weight for adults [8] and 1.6 (0.68 to 5.4) pg per kg body weight for small children [9] and the relatively long biological half-lives, the background values listed in Table 1 were determined [10]. As the concentrations are distinctly dependent on age, the total population sampled was divided into three approximately equal age groups. In addition, it has been observed that the background concentrations have shown a sharp drop in recent years and that this trend is still continuing at present – although at a slower rate [11].

Dioxins, furans and WHO PCB in whole blood

Table 1. Background levels for dioxins and furans in human blood in pg/g fat, Germany 1996 [10] (CDD = chlorodibenzodioxin, CDF = chlorodibenzofuran, n.d. = not detectable, 95% = 95th percentile, BGA = Bundesgesundheitsamt [German Ministry of Health], UBA = Umweltbundesamt [German Ministry of the Environment])

Age group No of samples	All age groups 18 to 71 years 180			Age group 1, 18 to 30 years 59			Age group 2, 31 to 42 years 68			Age group 3, 43 to 71 years 53		
pg/g blood fat	Median	Minimum	95%	Median	Minimum	95%	Median	Minimum	95%	Median	Minimum	95%
Age	36.0	18.0	55.0	26.0	18.0	30.0	37.0	31.0	42.0	47.0	43.0	58.0
2,3,7,8-Tetra-CDD	2.2	n.d.	4.2	1.9	1.0	3.7	2.3	n.d.	3.8	2.5	n.d.	4.8
1,2,3,7,8-Penta-CDD	5.4	1.7	9.5	4.1	1.7	8.0	5.6	1.9	8.7	6.6	3.3	10.3
1,2,3,4,7,8-Hexa-CDD	5.2	2.0	10.0	4.3	2.0	9.1	5.5	2.1	9.3	6.6	2.0	11.4
1,2,3,6,7,8-Hexa-CDD	20.1	3.7	40.6	15.6	3.7	30.1	24.2	5.7	40.8	25.8	6.8	44.4
1,2,3,4,7,8-Hexa-CDD	3.6	1.5	6.5	3.4	1.8	6.0	3.8	1.5	6.9	3.6	1.9	7.4
Sum of hexa-CDDs	29.4	8.4	55.1	23.8	8.4	44.7	33.2	11.6	56.5	37.8	14.9	58.9
1,2,3,4,6,7,8-Hepta-CDD	30.5	8.8	59.7	33.1	8.8	54.2	31.3	12.4	74.5	29.0	9.4	60.2
Octa-CDD	270.0	60.7	518.2	260.9	60.7	505.5	286.2	110.6	549.3	239.7	95.7	499.7
2,3,7,8-Tetra-CDF	1.2	0.5	1.9	1.2	0.5	1.8	1.3	0.5	2.0	1.1	0.5	1.9
1,2,3,7,8-Penta-CDF	0.5	n.d.	2.0	0.5	0.5	2.0	0.5	0.5	2.4	0.5	n.d.	1.0
2,3,4,7,8-Penta-CDF	10.1	3.2	20.1	7.8	3.2	13.8	10.3	3.9	20.1	13.9	5.2	24.9
Sum of penta-CDFs	10.2	3.8	20.6	7.9	3.8	14.4	10.4	3.9	20.1	14.3	5.2	24.9
1,2,3,4,7,8-Hexa-CDF	6.1	2.6	11.3	5.4	2.6	8.9	6.3	2.8	12.1	7.0	2.8	12.1
1,2,3,6,7,8-Hexa-CDF	4.6	2.0	8.4	3.9	2.2	6.2	4.8	2.0	8.3	5.2	2.2	8.8
2,3,4,6,7,8-Hexa-CDF	n.d.	n.d.	n.d.	n.d.	n.d.	n.d.	n.d.	n.d.	n.d.	n.d.	n.d.	n.d.
Sum of hexa-CDFs	2.4	0.5	3.7	2.4	1.4	3.7	2.4	0.5	3.8	2.4	1.0	3.7
1,2,3,4,6,7,8-Hepta-CDF	13.1	6.1	21.6	11.9	6.5	16.5	13.3	6.1	22.2	14.6	6.3	23.9
1,2,3,4,7,8,9-Hepta-CDF	7.7	3.1	15.0	8.9	4.2	13.8	7.6	4.7	16.7	7.0	3.1	13.9
Sum of hepta-CDFs	0.8	0.5	1.4	0.8	0.5	1.4	0.8	0.5	1.7	0.8	0.5	1.0
Octa-CDF	7.8	3.6	15.3	9.7	4.2	14.0	7.8	4.7	17.7	7.3	3.6	13.9
	2.5	1.5	2.5	2.5	1.8	2.5	2.5	1.6	2.5	2.5	1.5	2.5
Sum of penta-CDDs	336.4	85.5	613.4	321.4	85.5	591.7	358.5	159.1	686.3	331.6	128.1	606.8
Sum of penta-CDFs	35.4	19.7	56.4	34.3	19.7	48.1	35.6	19.7	56.2	39.6	19.7	61.3
Sum of penta-CDDs/penta-CDFs	374.4	107.3	653.8	352.8	107.3	628.8	405.4	188.2	733.7	368.1	148.3	653.7
I-TEQ (NATO/CCMS)	15.6	7.0	26.9	11.9	7.3	20.4	17.1	7.0	26.1	18.4	9.6	30.8
TEQ (BGA/UBA)	9.2	4.5	14.8	7.7	4.5	12.0	10.1	4.6	14.2	10.5	5.7	17.1

Whole blood is used as the principal investigation material worldwide, and the measured values are given in pg I-TEQ per g fat. This ensures that the measured results are comparable without further conversion, even for data from serum or plasma samples [12, 13], from fatty tissue [13] or from mother's milk [12]. Table 1 shows background values for persons in the general population that were determined in 1996.

As part of the reevaluation of the toxicity of PCDDs, PCDFs and PCBs, the WHO has introduced new toxicity equivalent factors [14] that have led to TEQ values approx. 15% higher than the I-TEQ values according to NATO/CCMS. Background levels for values according to this model are not available at present.

2,3,7,8-TCDD has been assigned to category 4 of the carcinogenic working materials, as it must be assumed that the dose-effect relationship for this substance with tumour-promoting properties is not linear ("effect threshold"). Due to the lack of data for humans on the exposure levels that lead to adverse effects caused by TCDD, the MAK value has been derived from the current background values. To date there is no indication that these background levels lead to relevant effects in humans. A toxicokinetic model was used to calculate that an intake of 10 pg TCDD/m^3 (MAK value) by inhalation for eight hours a day would not increase the background levels significantly [4, 15].

Author: *M. Ball*
Examiner: *J. Wittsiepe*

Dioxins, furans and WHO PCB in whole blood

Application Determination in whole blood

Analytical principle Capillary gas chromatography/
mass spectrometric detection (MS)

Completed in May 2001

Contents

1 General principles
2 Equipment, chemicals and solutions
2.1 Equipment
2.2 Chemicals
2.3 Solutions and conditioning
2.4 Calibration standards
3 Specimen collection and sample preparation
3.1 Specimen collection
3.2 Separation and gravimetric determination of the blood fats
3.3 Separation of the analytes from the fat fraction
4 Operational parameters
4.1 Operational parameters for gas chromatography and mass spectrometry
5 Analytical determination
6 Calibration
7 Calculation of the analytical results
8 Standardisation and quality control
9 Evaluation of the method
9.1 Precision
9.2 Accuracy
9.3 Detection limits
9.4 Sources of error
10 Discussion of the method
11 References
Addendum: Alternative fat extraction procedure

1 General principles

This method is based on a specific extraction of the blood fat from 40 g of whole blood, followed by a solid-phase clean-up of the extract on modified silica gels, active carbon and aluminium oxide [1]. The quantitative determination is carried out by means of capillary gas chromatography in combination with high-resolution mass spectrometry in the SIM mode. Calibration standards are prepared in toluene and are measured without further processing. The ^{13}C-labelled compounds corresponding to the analytes serve as internal standards. The toxicity equivalents based on 2,3,7,8-tetrachlorodibenzodioxin are calculated from the measured values with the help of the WHO calculation model.

2 Equipment, chemicals and solutions

2.1 Equipment

Gas chromatograph with high-resolution mass-selective detector, split-splitless injection system, autosampler and data processing system for evaluation

Capillary gas chromatographic column:
Length: 60 m; 0.25 mm; stationary phase: 5% phenyl/95% methylpolysiloxane; film thickness: 0.1 µm (e.g. J&W Scientific DB-5)

Precolumn:
Length and material: 5 m fused silica, inner diameter: 0.32 mm (e.g. from Restek, Art. No. 10044), transfer capillary (e.g. from Restek Art. No. 10043), universal capillary connector (e.g. from Analyt, Art. No. 20105 Q)

Rotary evaporator (e.g. Büchi RE 111) with regulated vacuum connection

Microcolumns containing active carbon (from ERGO, Hamburg, Germany)

Variable pipette, 0.1 to 2.0 mL (from Barkey)

Micropipettes (from Brand No. 708744)

Stirring rods made of stainless steel

Variable adjustable 100 µL and 1 mL pipettes (e.g. from Eppendorf)

All the glassware must be rinsed with toluene and n-hexane before use. It is important to ensure that the inner walls have been completely wetted with solvent while rinsing the vessels.

Glass chromatography columns without a frit or valve (specially made):

Column for blood (for separation and gravimetric determination of the fats from blood): Length = 260 mm, Diameter = 50 mm

Column for fat (to separate the fat fraction from the analytes):
Length = 240 mm, Diameter = 12 mm
Aluminium oxide type column (to separate the fat fraction from the analytes):
Length = 350 mm, Diameter = 8 mm

Tamper consisting of a Teflon stamp with a stainless steel shaft, designed to fit the chromatography column for blood

Active carbon microcolumn system (from Ergo)

1 L Glass column with frit and Teflon valve

1000 mL Round-bottomed flasks with hollow bulge to collect approx. 5 mL solution (specially made), glass stopper and Teflon cuffs

3000 mL Double-necked flasks, glass stoppers and Teflon cuffs

Reflux cooler for the 3000 mL double-necked flasks

Heatable magnetic stirrer with contact thermostat

Water bath or oil bath

25 mL Pear-shaped distilling flasks

50 mL Pear-shaped distilling flasks

Glass piston pipette, 10 mL (from Fortuna)

Glass Pasteur pipettes in various sizes

50 and 100 mL sample tubes

Sample vials for the autosampler with micro-inserts and aluminium crimp-caps

1, 5, 10, 50 and 100 mL volumetric flasks

2.2 Chemicals

Toluene for organic trace analysis (e.g. from Baker, Art. No. 8092)

Pentane for organic trace analysis (e.g. from Baker, Art. No. 6145)

n-Hexane for organic trace analysis (e.g. from Merck, Art. No. 4371)

Methanol for organic trace analysis (e.g. from Baker, Art. No. 5160)

Dichloromethane for organic trace analysis (e.g. from Baker, Art. No. 3023)

Ethanol for organic trace analysis (e.g. from Promochem, Art. No. 9063)

Isopropanol for organic trace analysis (e.g. from Baker, Art. No. 4359)

Sodium sulphate anhydrous, fine, pulverised (e.g. from Merck, Art. No. 106649)

Sodium sulphate anhydrous, coarse, granulate (e.g. from Merck, Art. No. 106603)

Active carbon, type AX 21 (manufacturer Amoco, available from Ergo, Hamburg) (Alternatively, Carbopack C active carbon (from Supelco, No. 1-0258) can be used, but the elution conditions must be adapted in this case.)

Sodium chloride p.a. (e.g. from Merck)

Native standard dioxin solution of all 17 dioxins and furans (e.g. CIL No. EDF 4175 1 µg/mL)

^{13}C-labelled standard dioxin solution of all 17 dioxins and furans (e.g. CIL No. EDF 4067 1 µg/mL)

Native standard solution of non-ortho-PCBs (e.g. CIL No. EC 4986, 1 µg/mL)

^{13}C-labelled standard solution of non-ortho-PCBs (e.g. CIL No. EC 4187, 1 µg/mL)

Native standard solution of mono-ortho-PCBs (e.g. CIL No. EC 4987, 10 µg/mL)

^{13}C-labelled standard solution of mono-ortho-PCBs (e.g. CIL No. EC 4188, 10 µg/mL)

^{13}C-labelled standard solution of 1,2,3,4-TCDD (e.g. CIL No. ED 911, 50 µg/mL)

Silanised glass wool (e.g. from Shimadzu, Art. No. 201-35311)

Glass fibre filter (e.g. from Whatman, diameter 55 mm, type GF/A, Art. No. 1821055)

Isolute HM – N (from IST Ltd., Art. No. 9800-5000)

Nitrogen 4.6 (e.g. from Westfalen Gas)

Caesium hydroxide (e.g. from Aldrich, Art. No. 19833-1)

Silica gel 60 (e.g. from Merck, Art. No. 107754)

Aluminium oxide (Aluminia B Super I, for dioxin analysis, from ICN, Art. No. 04569)

Sulphuric acid 95–98% (e.g. from Baker, Art. No. 6027)

Caesium silicate:
835 mL methanol are placed in a 3 L double-necked flask fitted with a reflux cooler. Then 250 g caesium hydroxide are added in portions while stirring (magnetic stirrer) (Take care! Strong generation of heat), and the mixture is allowed to cool to room temperature while stirring. The solution is filtered into a further 3 L double-necked flask through a glass fibre filter to remove undissolved caesium hydroxide. Then 420 g silica gel and an additional 835 mL methanol are added. The mixture is heated to 55 °C for two hours while stirring in the oil bath (reflux cooler). After cooling, the precipitated caesium silicate is transferred to a 1 L glass column with a frit. The caesium silicate is washed with 1 L methanol and 3 L dichloromethane and subsequently dried in a stream of nitrogen. The caesium silicate is stored at room temperature in sealed glass vessels. It is stable for 6 months.

Silica gel treated with sulphuric acid:
240 g purified silica gel (extracted with hexane, toluene and dichloromethane) are weighed into a 1000 mL round-bottomed flask. 160 g sulphuric acid are slowly added and the round-bottomed flask is sealed with a glass stopper. The flask is shaken intensively for 1 minute (Take care! Strong generation of heat) and the flask is subsequently ventilated. This shaking procedure must be repeated at least 6 times over a period of 16 hours. The silica gel treated with sulphuric acid is stored at room temperature in the round-bottomed flask used for mixing. The stability of the silica gel thus treated with sulphuric acid is 3 weeks.

Activated aluminium oxide:
Aluminium oxide is filled into a glass beaker and activated at 180 °C for at least 48 hours in the drying cupboard. It is also stored there.

2.3 Solutions and conditioning

Injection standard (10 µg 1,2,3,4-^{13}C-TCDD / L toluene)

20 µL of the commercially available standard solution 1,2,3,4-TCDD (^{13}C-labelled, 50 mg/L) are placed in a 100 mL volumetric flask. The volumetric flask is then filled to its nominal volume with toluene.
The injection standard is used to dissolve the analytes isolated from the matrix (Section 3.3) and to inject them into the gas chromatograph. The injection standard serves to check the absolute recovery.

Isopropanol/n-hexane (2:3, v/v)

Approx. 640 mL of a mixture of 2 parts of isopropanol and 3 parts of n-hexane are required for each sample to be processed.

n-Hexane/dichloromethane (98:2, v/v)

7 mL of a mixture of 98 parts of n-hexane and 2 parts of dichloromethane are required for each sample to be processed.

n-Hexane/dichloromethane (1:1, v/v)

25 mL of a mixture of 1 part of n-hexane and 1 part of dichloromethane are required for each sample to be processed.

Chromatography columns for extraction of fats from blood

The chromatography columns for blood are filled with dry Isolute and sodium chloride in the following sequence:

Glass fibre filter
Isolute (20 g)
Glass fibre filter
Sodium chloride (10 g)
Isolute (25 g)
Sodium chloride (10 g)
Isolute (25 g)
Sodium chloride (10 g)

The columns thus filled are rinsed with 250 mL isopropanol/n-hexane (2:3 v/v) and then with 250 mL dichloromethane. If necessary, this process can be accelerated by using a stream of slightly compressed nitrogen. Then the solvent is removed as completely as possible by purging with nitrogen, and the columns are subsequently stored for at least 10 hours at 60 °C in the drying cupboard.

Active carbon microcolumn system to separate blood fats

Before use, the active carbon microcolumn system (Figure 1) must be cleaned as shown in Table 2. For this purpose, the direction of flow is set on the instrument and the relevant solvents are allowed to flow through the system. 50 mg AX21 active carbon are used for each system, the column systems can be used repeatedly.

Table 2. Cleaning the active carbon microcolumn system

Solvent	Direction of flow
30 mL Toluene	Reverse flow
20 mL Methanol	Reverse flow
20 mL Toluene	Forward flow
20 mL Toluene	Reverse flow
20 mL n-Hexane	Reverse flow
5 mL n-Hexane	Forward flow

Mixed column (column for fat) for cleaning up the extract

A glass column (column for fat) is filled alternately with 0.75 g silica gel treated with sulphuric acid and 0.35 g caesium silicate and then washed with 15 mL n-hexane. These columns must be freshly prepared on the day of the analysis.

Column (aluminium oxide type) for cleaning up the extract

A glass column (aluminium oxide type) is filled with 3.65 g hot, activated aluminium oxide, and approx. 0.5 cm fine sodium sulphate is subsequently overlaid. These columns must be freshly prepared on the day of the analysis.

2.4 Calibration standards

2.4.1 Internal standard

Spiking solution for the internal standard:
100 µL of each of the ^{13}C-labelled dioxins and furans, the ^{13}C-labelled non-ortho-PCBs and the ^{13}C-labelled mono-ortho-PCBs are pipetted into a 5 mL volumetric flask. The flask is subsequently filled to its nominal volume with toluene. The concentrations of the internal standards in this solution are shown in Table 3.

Table 3. Concentrations of the ^{13}C-labelled internal standards in the spiking solution

Substance	Concentration [mg/L]
3,3′,4,4′-Tetrachlorobiphenyl (PCB 77)	0.02
3,4,4′,5-Tetrachlorobiphenyl (PCB 81)	0.02
3,3′,4,4′,5-Pentachlorobiphenyl (PCB 126)	0.02
3,3′,4,4′,5,5′-Hexachlorobiphenyl (PCB 169)	0.02
2,3,3′,4,4′-Pentachlorobiphenyl (PCB 105)	0.2
2,3,4,4′,5-Pentachlorobiphenyl (PCB 114)	0.2
2,3′,4,4′,5-Pentachlorobiphenyl (PCB 118)	0.2
2′,3,4,4′,5-Pentachlorobiphenyl (PCB 123)	0.2
2,3,3′,4,4′,5-Hexachlorobiphenyl (PCB 156)	0.2
2,3,3′,4,4′,5′-Hexachlorobiphenyl (PCB 157)	0.2
2,3′,4,4′,5,5′-Hexachlorobiphenyl (PCB 167)	0.2
2,3,3′,4,4′,5,5′-Heptachlorobiphenyl (PCB 189)	0.2
2,3,7,8-TCDD	0.02
1,2,3,7,8-PentaCDD	0.02
1,2,3,4,7,8-HexaCDD	0.02
1,2,3,6,7,8-HexaCDD	0.02
1,2,3,7,8,9-HexaCDD	0.02
1,2,3,4,6,7,8-HeptaCDD	0.02
OctaCDD	0.02
2,3,7,8-TCDF	0.02
1,2,3,7,8-PentaCDF	0.02
2,3,4,7,8-PentaCDF	0.02
1,2,3,4,7,8-HexaCDF	0.02
1,2,3,6,7,8-HexaCDF	0.02
1,2,3,7,8,9-HexaCDF	0.02
2,3,4,6,7,8-HexaCDF	0.02
1,2,3,4,6,7,8-HeptaCDF	0.02
1,2,3,4,7,8,9-HeptaCDF	0.02
OctaCDF	0.02

2.4.2 Calibration standards

Starting solution:
100 µL of each of the native dioxins and furans, the non-*ortho*-PCBs and the mono-ortho-PCBs are pipetted into a 1 mL volumetric flask. The volumetric flask is subsequently filled to its nominal volume with toluene. This solution contains 0.1 mg/L of each component, with the exception of the mono-ortho-PCBs, which are present in a concentration of 1 mg/L.

Table 4. Concentrations of the calibration standards (I.S. = internal standard; CB = chlorobiphenyl)

	Calibration standard 1	Calibration standard 2	Calibration standard 3	Calibration standard 4	Calibration standard 5
	100 µL of the spiking solution of the internal standard in each case				
	4 µL starting solution	10 µL starting solution	20 µL starting solution	50 µL starting solution	100 µL starting solution
	mg/L	mg/L	mg/L	mg/L	mg/L
2,3,7,8-TCDD	0.0004	0.001	0.002	0.005	0.01
1,2,3,7,8-PentaCDD	0.0004	0.001	0.002	0.005	0.01
1,2,3,4,7,8-HexaCDD	0.0004	0.001	0.002	0.005	0.01
1,2,3,6,7,8-HexaCDD	0.0004	0.001	0.002	0.005	0.01
1,2,3,7,8,9-HexaCDD	0.0004	0.001	0.002	0.005	0.01
1,2,3,4,6,7,8-HeptaCDD	0.0004	0.001	0.002	0.005	0.01
OctaCDD	0.0004	0.001	0.002	0.005	0.01
1,2,7,8-TCDF	0.0004	0.001	0.002	0.005	0.01
2,3,7,8-TCDF	0.0004	0.001	0.002	0.005	0.01
1,2,3,7,8-PentaCDF	0.0004	0.001	0.002	0.005	0.01
2,3,4,7,8-PentaCDF	0.0004	0.001	0.002	0.005	0.01
1,2,3,4,7,8-HexaCDF	0.0004	0.001	0.002	0.005	0.01
1,2,3,6,7,8-HexaCDF	0.0004	0.001	0.002	0.005	0.01
1,2,3,7,8,9-HexaCDF	0.0004	0.001	0.002	0.005	0.01
2,3,4,6,7,8-HexaCDF	0.0004	0.001	0.002	0.005	0.01
1,2,3,4,6,7,8-HeptaCDF	0.0004	0.001	0.002	0.005	0.01
1,2,3,4,7,8,9-HeptaCDF	0.0004	0.001	0.002	0.005	0.01
OctaCDF	0.0004	0.001	0.002	0.005	0.01
3,4,4',5-TetraCB (81)	0.0004	0.001	0.002	0.005	0.01
3,3',4,4',5-PentaCB (126)	0.0004	0.001	0.002	0.005	0.01
3,3',4,4',5,5'-HexaCB (169)	0.0004	0.001	0.002	0.005	0.01
2,3,3',4,4'-PentaCB (105)	0.004	0.01	0.02	0.05	0.1
2,3,4,4',5-PentaCB (114)	0.004	0.01	0.02	0.05	0.1
2,3',4,4',5-PentaCB (118)	0.004	0.01	0.02	0.05	0.1
2',3,4,4',5-PentaCB (123)	0.004	0.01	0.02	0.05	0.1
2,3,3',4,4',5-HexaCB (156)	0.004	0.01	0.02	0.05	0.1
2,3,3',4,4',5'-HexaCB (157)	0.004	0.01	0.02	0.05	0.1
2,3',4,4',5,5'-HexaCB (167)	0.004	0.01	0.02	0.05	0.1
2,3,3',4,4',5,5'-HeptaCB (189)	0.004	0.01	0.02	0.05	0.1

Calibration standards containing concentrations of the substances between 400 ng/L and 10,000 ng/L are prepared from the starting solution by dilution with toluene and addition of the internal standards.

For this purpose the appropriate volumes of the working solution are pipetted into a 1 mL volumetric flask in accordance with the pipetting scheme shown in Table 4. In addition, 100 µL of the spiking solution of the internal standard are added to each flask using a pipette. The volumetric flasks are subsequently filled to their nominal volume with toluene.

The calibration standards are stable for at least 1 year when stored at approximately −18 °C.

3 Specimen collection and sample preparation

3.1 Specimen collection

A total of approx. 45 mL blood are withdrawn from the arm vein using several EDTA Monovettes. After shaking intensively, the contents of the Monovettes are combined in a 100 mL sample vessel and stored at approx. −18 °C until sample processing can be carried out. The samples can be stored for several years, experience with storage of pooled samples has been gathered over a period of 4 years.

3.2 Separation and gravimetric determination of the blood fats

Approx. 40.0 g blood and 30.0 g water are weighed exactly into a 100 mL sample vessel one after another. Then 5 mL ethanol and 20 µL of the spiking solution of the internal standard are added to the mixture. The sample vessels are sealed and shaken intensively for 5 minutes on a laboratory shaker.

The contents of the sample vessel are completely transferred onto the prepared extraction columns for blood. The upper salt layer is stirred with a stirring rod during transfer. The sample vessel is rinsed four times with 100 mL isopropanol/n-hexane in each case. Then 200 mL isopropanol/n-hexane are applied to the column, and the solvent is forced out of the column by nitrogen. The eluate (approx. 600 mL) is collected in a 1000 mL round-bottomed flask and evaporated on a rotary evaporator until a few millilitres remain. This is achieved on a water bath at a temperature of 40 °C with an initial vacuum of approx. 400 mbar. The pressure is then lowered to 200 mbar and finally to 45 mbar. The residue is transferred using pentane to a dry 25 mL pear-shaped distilling flask that has been previously weighed exactly to 0.1 mg. In order to remove any accompanying water, the extract is previously passed through a chromatography column suitable for fat filled with 4 g sodium sulphate. The round-bottomed flask is rinsed three times with 4 mL pentane in each case. After all the extracts have passed through the drying column, the column outlet is rinsed with a little pentane.

The solvent is subsequently evaporated off on a water bath at a temperature of 40 °C and a pressure of 600 to 40 mbar until the weight remains constant. This procedure

usually requires several weighing steps. The quantity of fat is normally between 2 and 7 g per litre blood.
The fat content of the sample is calculated according to equation 1:

$$m_{\text{Blood fat}} = \frac{m_{\text{Fat}}}{m_{\text{Blood}}} \cdot 100 \tag{1}$$

where:

$m_{\text{Blood fat}}$	Mass of fat in 100 g whole blood [g/100 g]
m_{Fat}	Final weighed amount of fat [g]
m_{Blood}	Weighed-in amount of whole blood sample [g].

3.3 Separation of the analytes from the fat fraction

The clean-up of the fat fraction is carried out by separation of the analytes on a special active carbon (AX21) column, followed by a mixed column, an aluminium oxide column and a further mixed column.

The extracted fat is dissolved in 6 mL n-hexane in a pear-shaped distilling flask, and 2 g silica gel treated with sulphuric acid are added. After being mixed thoroughly, the suspension is transferred to the sample reservoir of the active carbon microcolumn system (see Section 2.3). The pear-shaped distilling flask is rinsed 3 times with 3 mL n-hexane in each case to ensure complete transfer. The active carbon system is set at *forward flow,* and the sample is flushed through the system with 2×10 mL and 1×20 mL n-hexane. Then the system is set for *reverse flow* and the planar molecules are eluted from the active carbon column using 30 mL toluene. The eluate is collected in a 50 mL pear-shaped distilling flask and evaporated at 40 °C and 70 mbar in the rotary evaporator until approx. 1 mL remains.

The complete quantity of the sample is then transferred to a mixed column for suitable fat (see Section 2.3: this column contains caesium silicate and silica gel treated with sulphuric acid). The pear-shaped distilling flask is rinsed with 2×4 mL n-hexane to ensure complete transfer. The eluate from the mixed column is directly transferred to the prepared aluminium oxide column (see Section 2.3: this column contains aluminium oxide overlaid with sodium sulphate), and 7 mL n-hexane are used for rinsing. The aluminium oxide column is washed with 6 mL n-hexane/dichloromethane (98:2 v/v), the washing liquid is discarded and the column is subsequently eluted with 25 mL n-hexane/dichloromethane (1:1 v/v) into a 50 mL pear-shaped distilling flask. The eluate is evaporated at 330 to 250 mbar and 40 °C on the rotary evaporator until 0.5 mL remains, then it is transferred with 2 mL n-hexane to the second mixed column. After elution with 3×2 mL n-hexane, the eluate is evaporated almost to dryness at 330 to 250 mbar and 40 °C on the rotary evaporator.

The sample residue is dissolved in n-hexane (0.2 mL), transferred to the micro-insert of the sample vial for the autosampler using a micropipette and then dried in a stream of nitrogen. The sample is subsequently redissloved by addition of 10 µL of the solution of the injection standard. This solution is injected into the GC.

4 Operational parameters

4.1 Operational parameters for gas chromatography and mass spectrometry

Capillary column:	Material:	Fused silica
	Stationary phase:	DB-5
	Length:	60 m
	Inner diameter:	0.25 mm
	Film thickness:	0.1 µm
Precolumn:	Material:	Fused silica
	Stationary phase:	none
	Length:	5 m
	Inner diameter:	0.32 mm
Detector:	High-resolution VG AutoSpec mass spectrometer	
Temperatures:	Column:	Initial temperature 90 °C, isothermal for 3 minutes, then increase at a rate of 25 °C/min to 210 °C, increase at a rate of 3 °C/min to 275 °C, then 20 min at the final temperature
	Injector:	270 °C
	Transfer line:	280 °C
Carrier gas:	Helium 4.6 at a pre-pressure of 27 psi (185 kPa)	
Split:	Splitless, split on after 180 s	
Sample volume:	2 µL	
Ionisation type:	Electron impact ionisation (EI)	
Ionisation energy:	29 eV	
Dwell time:	See Table 5	

All other parameters must be optimised in accordance with the manufacturer's instructions.

5 Analytical determination

In each case 2 µL of the blood samples processed as described in Section 3 are injected into the gas chromatograph for the analytical determination. A quality control sample and an aqueous blank sample are analysed with each analytical series. The fragments shown in Table 5 are recorded in the SIM mode. The retention times of the individual substances are given in Table 6. On principle, both masses are used for evaluation. If differences of more than 25% are found, the cause must be sought and the mass without interference must be used for evaluation.

Table 5. Masses recorded for evaluation

Analyte	Mass [m/z]	Dwell time [ms]	Delay [ms]
Group 1			
Tetrachlorobiphenyls	289.9224	60	20
Tetrachlorobiphenyls	291.9194	60	10
Tetrachlorobiphenyl-$^{13}C_{12}$	301.9626	40	10
Tetrachlorobiphenyl-$^{13}C_{12}$	303.9597	40	10
TCDF	303.9016	60	10
TCDF	305.8987	60	10
TCDF-$^{13}C_{12}$	315.9418	40	10
TCDF-$^{13}C_{12}$	317.9389	40	10
TCDD	319.8965	60	10
TCDD	321.8936	60	10
Pentachlorobiphenyls	325.8805	60	10
Pentachlorobiphenyls	327.8775	60	10
Lock mass check	330.9792	20	10
Lock mass	330.9792	50	10
TCDF-$^{13}C_{12}$	331.9368	40	10
TCDF-$^{13}C_{12}$	333.9339	40	10
Pentachlorobiphenyl-$^{13}C_{12}$	337.9207	40	10
Pentachlorobiphenyl-$^{13}C_{12}$	339.9177	40	10
Group 2			
PentaCDF	339.8597	60	20
PentaCDF	341.8568	60	10
Lock mass check	342.9793	20	10
Lock mass	342.9793	50	10
PentaCDF-$^{13}C_{12}$	351.9000	40	10
PentaCDD	353.8576	60	10
PentaCDF-$^{13}C_{12}$	353.8970	40	10
PentaCDD	355.8546	60	10
Hexachlorobiphenyls	359.8415	60	10
Hexachlorobiphenyls	361.8385	60	10
PentaCDF-$^{13}C_{12}$	365.8978	40	10
PentaCDF-$^{13}C_{12}$	367.8949	40	10
Hexachlorobiphenyl-$^{13}C_{12}$	371.8817	40	10
Hexachlorobiphenyl-$^{13}C_{12}$	373.8788	40	10

Table 5 (continued)

Analyte	Mass [m/z]	Dwell time [ms]	Delay [ms]
Group 3			
HexaCDF	373.8207	60	20
HexaCDF	375.8178	60	10
Lock mass check	380.9760	20	10
Lock mass	380.9760	50	10
HexaCDF-$^{13}C_{12}$	385.8610	40	10
HexaCDF-$^{13}C_{12}$	387.8581	40	10
HexaCDD	389.8156	60	10
HexaCDD	391.8127	60	10
Heptachlorobiphenyls	393.8025	60	10
Heptachlorobiphenyls	395.7996	60	10
HexaCDF-$^{13}C_{12}$	401.8559	40	10
HexaCDF-$^{13}C_{12}$	403.8530	40	10
Heptachlorobiphenyls-$^{13}C_{12}$	405.8428	40	10
Heptachlorobiphenyls-$^{13}C_{12}$	407.8398	40	10
Group 4			
HeptaCDF	407.7818	60	20
HeptaCDF	409.7788	60	10
HeptaCDF-$^{13}C_6$	415.7989	40	10
HeptaCDF-$^{13}C_6$	417.7959	40	10
HeptaCDF-$^{13}C_{12}$	419.8220	40	10
HeptaCDF-$^{13}C_{12}$	421.8191	40	10
HeptaCDD	423.7767	60	10
HeptaCDD	425.7737	60	10
Lock mass check	430.9728	20	10
Lock mass	430.9728	50	10
HeptaCDF-$^{13}C_{12}$	435.8169	40	10
HeptaCDF-$^{13}C_{12}$	437.8140	40	10
Group 5			
OctaCDF	441.7428	60	20
OctaCDF	443.7398	60	10
OctaCDF-$^{13}C_{12}$	453.7831	40	10
Lock mass check	454.9728	20	10
Lock mass	454.9728	50	10
OctaCDF-$^{13}C_{12}$	455.7801	40	10
OctaCDD	457.7377	60	10
OctaCDD	459.7348	60	10
OctaCDD-$^{13}C_{12}$	469.7780	40	10
OctaCDD-$^{13}C_{12}$	471.7750	40	10

Table 6. Retention times

Analyte	Retention time (min)		Analyte*	Retention time (min)	
	Analyte	^{13}C-IS		Analyte	^{13}C-IS
2,3,7,8-TCDF	20.15	20.13	PCB 81	17.09	17.07
1,2,3,4-TCDD	19.97	19.97	PCB 77	17.39	17.37
2,3,7,8-TCDD	19.66	19.65	PCB 123	18.10	18.08
1,2,3,7,8-PentaCDF	22.90	22.88	PCB 118	18.18	18.16
2,3,4,7,8-PentaCDF	23.63	23.60	PCB 114	18.57	18.56
1,2,3,7,8-PentaCDD	23.98	23.95	PCB 105	19.10	19.10
1,2,3,4,7,8-HexaCDF	26.86	26.86	PCB 126	20.36	20.35
1,2,3,6,7,8-HexaCDF	26.99	26.79	PCB 167	21.11	21.10
1,2,3,7,8,9-HexaCDF	28.64	28.58	PCB 156	22.07	22.05
2,3,4,6,7,8-HexaCDF	27.63	27.63	PCB 157	22.30	22.29
1,2,3,4,7,8-HexaCDD	27.82	27.89	PCB 169	23.73	23.74
1,2,3,6,7,8-HexaCDD	27.93	27.91	PCB 189	25.39	25.38
1,2,3,7,8,9-HexaCDD	28.33	28.31			
1,2,3,4,6,7,8-HeptaCDF	30.90	30.87			
1,2,3,4,7,8,9-HeptaCDF	33.17	33.16			
1,2,3,4,6,7,8-HeptaCDD	32.55	32.54			
OctaCDF	38.09	38.07			
OctaCDD	37.93	37.91			

* See also Table 7 for the numerical designation of the polychlorinated biphenyls.

The retention times shown in Table 6 serve only as a guide. Users of the method must satisfy themselves of the separation power of the capillary column used and the resulting retention behaviour of the substances. An example showing a chromatogram of a processed human blood sample is given in Figure 2.

6 Calibration

The calibration standards are analysed without further treatment by means of gas chromatography/mass spectrometry in accordance with Sections 4 and 5. In each case 2 µL of the standards are injected into the gas chromatograph. Calibration graphs are obtaining by plotting the quotients of the peak area of the analytes and the relevant internal standard as a function of the absolute quantities of the calibration standards (mass of the analyte in 2 µL toluene). It is not necessary to plot a complete calibration graph for every analytical series. It is sufficient to analyse one calibration standard for every analytical series. The ratio of the value obtained for this standard and the result for the equivalent standard in the complete calibration graph is calculated. Using this quotient each result read off the calibration graph is adjusted for the relevant series.
The injection standard serves to calculate the absolute recovery. This value indicates possible problems during processing, but it is not used for calculating the result.
New calibration graphs should be plotted if the quality control results indicate systematic deviation.
The calibration graph is linear between the detection limit and 500 µg per litre toluene.

7 Calculation of the analytical results

Quotients are calculated by dividing the peak areas of the analytes by that of the relevant internal standard. These quotients are used to read off the corresponding concentration of the analytes in pg per weighed-in quantity of blood (40 g) from the relevant calibration graph.

Any reagent blank values must be subtracted from the analytical results for the real samples.

The result is then divided by the weight of blood fat (see Section 3.2) to give the result in pg/g of blood fat.

The dioxin, furan and PCB concentrations in blood fat are multiplied by the WHO toxicity equivalent factors (WHO TEF, Table 7) and the resulting products are added. This gives a WHO toxicity equivalent (WHO TEQ) in pg/g of fat.

It is also meaningful to express the result in pg per kilogram or pg per litre of whole blood in addition to the value in pg per g of blood fat. Experience has shown prob-

Table 7. WHO toxicity equivalent factors (WHO TEF)

Analyte	WHO TEF
2,3,7,8-TCDD	1
1,2,3,7,8-PentaCDD	1
1,2,3,4,7,8-HexaCDD	0.1
1,2,3,6,7,8-HexaCDD	0.1
1,2,3,7,8,9-HexaCDD	0.1
1,2,3,4,6,7,8-HeptaCDD	0.01
OctaCDD	0.0001
2,3,7,8-TCDF	0.1
1,2,3,7,8-PentaCDF	0.05
2,3,4,7,8-PentaCDF	0.5
1,2,3,4,7,8-HexaCDF	0.1
1,2,3,6,7,8-HexaCDF	0.1
1,2,3,7,8,9-HexaCDF	0.1
2,3,4,6,7,8-HexaCDF	0.1
1,2,3,4,6,7,8-HeptaCDF	0.01
1,2,3,4,7,8,9-HeptaCDF	0.01
OctaCDF	0.0001
3,4,4',5-TCB (PCB 81)	0.0001
3,3',4,4'-TCB (PCB 77)	0.0001
3,3',4,4',5-PentaCB (PCB 126)	0.1
3,3',4,4',5,5'-HexaCB (PCB 169)	0.01
2,3,3',4,4'-PentaCB (PCB 105)	0.0001
2,3,4,4',5-PentaCB (PCB 114)	0.0005
2,3',4,4',5-PentaCB (PCB 118)	0.0001
2',3,4,4',5-PentaCB (PCB 123)	0.0001
2,3,3',4,4',5-HexaCB (PCB 156)	0.0005
2,3,3',4,4',5'-HexaCB (PCB 157)	0.0005
2,3',4,4',5,5'-HexaCB (PCB 167)	0.00001
2,3,3',4,4',5,5'-HeptaCB (PCB 189)	0.0001

lems in the determination of blood fat can be recognised, and difficulties in comparing results with those of external investigations can be avoided.

8 Standardisation and quality control

Quality control of the analytical results is carried out as stipulated in the guidelines of the Bundesärztekammer (German Medical Association) [16, 17] and in the special preliminary remarks to this series. A native human blood sample is included in the analysis for the purpose of precision control. As a relatively large quantity of control material (40 g) is required for each analysis, it is advisable to use whole blood from a blood bank (if necessary, after its shelf-life has expired). Whole blood is pooled for this purpose. A year's supply of this control material is divided into aliquots in sealable 50 mL glass vessels and stored in the deep-freezer. The theoretical value and the tolerance range for this quality control material are determined in the course of a pre-analytical period (one analysis of the control material on each of 6 different days) [18–20].

External quality control can be achieved by participation in round-robin experiments. For example, the WHO offers round-robin experiments at irregular intervals.

9 Evaluation of the method

9.1 Precision

A sample of pooled human whole blood was processed and analysed several times on one day to determine the precision in the series. Six replicate determinations of this blood sample yielded the precision in the series documented in Table 8.

Table 8. Precision in the series (n = 6)

Parameter	Concentration [pg/g]	Standard deviation (rel.) [%]	Prognostic range [%]
2,3,7,8-TCDD	2.4	9.5	23.7
1,2,3,6,7,8-HexaCDD	28	3.9	9.7
2,3,4,7,8-PentaCDF	13	4.7	11.7
Σ 2,3,7,8 PCDD and PCDF (WHO TEQ)	20	3.5	8.7
3,3',4,4',5-PentaCB (126)	71	1.9	4.7
2,3',4,4',5-PentaCB (118)	18000	3.2	8.0

In addition, the precision from day to day was determined. A further sample of pooled human whole blood was processed and analysed on different days for this purpose. The precision results are shown in Table 9.

Table 9. Precision from day to day

Parameter	n	Concentration [pg/g]	Standard deviation (rel.) [%]	Prognostic range [%]
2,3,7,8-TCDD	16	2.0	22.7	48.1
1,2,3,6,7,8-HexaCDD	17	19	19.9	42.2
2,3,4,7,8-PentaCDF	17	12	15.5	32.8
Σ 2,3,7,8 PCDD and PCDF (WHO TEQ)	17	18	13.9	29.4
3,3',4,4',5-PentaCB (126)	17	71	25.6	54.2
2,3',4,4',5-PentaCB (118)	17	13000	11.8	25.0

9.2 Accuracy

The relative recovery rate was 100% in each case due to the use of the relevant isotope-labelled internal standards. The accuracy of the method was evaluated objectively by successful participation in the 2nd and 3rd round-robin experiments of the WHO.

9.3 Detection limits

Under the conditions described here the detection limits, calculated as three times the signal/noise ratio of the analytical background interference in the temporal environment of the analyte signals, were between 1 and 100 pg/g blood fat. The detection limits for the parameters determined using this method are shown in Table 10.

Table 10. Detection limits in pg/g blood fat

Analyte	Detection limit [pg/g]
2,3,7,8-TCDD	1
1,2,3,7,8-PentaCDD	1
1,2,3,4,7,8-HexaCDD	1
1,2,3,6,7,8-HexaCDD	1
1,2,3,7,8,9-HexaCDD	1
1,2,3,4,6,7,8-HeptaCDD	10
OctaCDD	30
2,3,7,8-TCDF	2
1,2,3,7,8-PentaCDF	1
2,3,4,7,8-PentaCDF	1
1,2,3,4,7,8-HexaCDF	2
1,2,3,6,7,8-HexaCDF	2
1,2,3,7,8,9-HexaCDF	2
2,3,4,6,7,8-HexaCDF	2
1,2,3,4,6,7,8-HeptaCDF	10

Table 10 (continued)

Analyte	Detection limit [pg/g]
1,2,3,4,7,8,9-HeptaCDF	5
OctaCDF	50
3,4,4',5-TCB (PCB 81)	5
3,3',4,4'-TCB (PCB 77)	100
3,3',4,4',5-PentaCB (PCB 126)	10
3,3',4,4',5,5'-HexaCB (PCB 169)	2
2,3,3',4,4'-PentaCB (PCB 105)	10
2,3,4,4',5-PentaCB (PCB 114)	10
2,3',4,4',5-PentaCB (PCB 118)	10
2',3,4,4',5-PentaCB (PCB 123)	100
2,3,3',4,4',5-HexaCB (PCB 156)	10
2,3,3',4,4',5'-HexaCB (PCB 157)	10
2,3',4,4',5,5'-HexaCB (PCB 167)	10
2,3,3',4,4',5,5'-HeptaCB (PCB 189)	10

9.4 Sources of error

Interference is hardly observed on account of the very specific chromatographic clean-up, including the use of active carbon. In particular, interference caused by polychlorinated biphenyls, which cannot be eliminated by mass spectrometric detection, is avoided because these substances are completely separated by the use of the special active carbon.

If extremely high levels of fat are present (>10 g/L blood), selective loss of mono-ortho-PCB can occur. The losses due to processing can be approx. 75% in such cases (absolute recovery approx. 25%).

10 Discussion of the method

These analytical instructions describe an efficient tool for determination of the representatives of the dibenzodioxins, dibenzofurans and dioxin-like PCBs that are classified as important in the current evaluation scheme of the WHO. On account of its sensitivity the method is suitable for reliably detecting the background levels of these persistent organochlorine compounds in the general population.

One particular advantage of this method is that it is especially suitable for relatively small sample quantities. The weighed-in sample amount can be reduced to approx. 20 g whole blood (10 g plasma). The greatest problem in this case is not the potential lack of sensitivity of the instrument but the relatively high blank value on account of the ubiquitous presence of the substances to be measured.

One possibility of lowering the blank value is to reduce the amount of solvent used. An alternative procedure for the isolation of blood fats is described in the addendum to this chapter. The two extraction methods have proved equivalent in international round-robin experiments. However, the reliability criteria given here were determined with the extraction procedure described in Section 3.

The high cost of the required ^{13}C-labelled standard substances is one of the drawbacks of all the analytical methods in this field. Therefore it is only practicable to adopt this method if the number of samples to be analysed is relatively large.

Instruments used:
Hewlett-Packard HP5890 gas chromatograph with Autospec VG mass selective detector, with a split-splitless injection system and Hewlett-Packard HP7673 autosampler, computer controlled system, Digital Vax Station 4000 60, program: Opus.

11 References

[1] *L.M. Smith, D.L. Stalling* and *J.L. Johnson:* Determination of part-per-trillion levels of polychlorinated dibenzofurans and dioxins in environmental samples. Anal. Chem. 56, 1830 (1984)

[2] *K. Rozmann, B. Stahl* and *H. Sterzl-Eckert:* Polychlorierte Dibenzodioxine und Dibenzofurane. In: *H. Greim* and *E. Deml (eds.)*: Toxikologie: Eine Einführung für Naturwissenschaftler und Mediziner. Verlag Wiley-VCH, Weinheim (1996)

[3] *H. Sagunski* and *G. Perger:* Biozide. In: *H. Marquardt* and *S.G. Schäfer (eds.)*: Lehrbuch der Toxikologie. Spektrum Verlag, Heidelberg (1997)

[4] *H. Greim (ed.):* 2,3,7,8-Tetrachlordibenzo-p-dioxin. Toxikologisch-arbeitsmedizinische Begründung von MAK-Werten, 28th issue. Wiley-VCH, Weinheim (1999)

[5] *H. Beck, A. Droß* and *W. Mathar:* PCDDs, PCDFs and related contaminants in the German food supply. Chemosphere 25, 1539–1550 (1992)

[6] *P. Fürst, C. Fürst* and *W. Groebel:* Levels of PCDDs and PCDFs in Food-Stuffs from the Federal Republic of Germany. Chemosphere 20, 787–792 (1990)

[7] 2nd Report of the Bund/Länder-Arbeitsgruppe DIOXINE 1992/1993 (1993)

[8] *P. Schrey, J. Wittsiepe, P. Mackrodt* and *F. Selenka:* Human fecal PCDD/F-excretion exceeds the dietary intake. Chemosphere 37 (9–12), 1825–1831 (1998)

[9] *J. Wittsiepe, P. Schrey* and *M. Wilhelm:* Dietary intake of PCDD/F by small children with different food consumption measured by the duplicate method. Chemosphere 43, 881–887 (2001)

[10] *O. Päpke, T. Herrmann* and *B. Schilling:* PCDD/Fs in Humans, Follow up of Background Data for Germany 1988/99. Organohalogen Compounds, Vol. 44, 221–224 (1999)

[11] *J. Wittsiepe, P. Schrey, U. Ewers, F. Selenka* and *M. Wilhelm:* Decrease of PCDD/F levels in human blood from Germany over the past ten years (1989–1998). Chemosphere 40, 1103–1109 (2000)

[12] *O. Päpke:* Aufbau der Dioxinbelastung des Menschen und Expositionsmessung in verschiedenen Matrizes. Arbeitsmed. Sozialmed. Umweltmed. Sonderheft 24, 11–23 (1998)

[13] *A. Schecter, J.J. Ryan, O. Päpke* and *M. Ball:* Comparison of Dioxin and Dibenzofuran Levels in Whole Blood, Blood Plasma and Adipose Tissue on a Lipid Basis. Chemosphere 23, 1913–1919 (1991)

[14] *M. van den Berg, L. Birnbaum, B.T.C. Bosveld, B. Brunström, P. Cook, M. Feeley, J.P. Giesy, A. Hanberg, R. Hasegawa, S.W. Kennedy, T. Kubiak, J.C. Larsen, F.X.R. van Leeuwen, A.K.D. Liem, C. Nolt, R.E. Petersen, L. Poellinger, S. Safe, D. Schrenck, D. Tillit, M. Tysklind, M. Younes, F. Wærn* and *T. Zacharewski:* Toxic Equivalency Factors (TEFs) for PCBs, PCDDs, PCDFs for Humans and for Wildlife. Environ. Health Perspect. 106(12), 775 (1998)

[15] *Deutsche Forschungsgemeinschaft:* MAK- und BAT-Werte-Liste, Report 37. Wiley-VCH, Weinheim (2001)

[16] *Bundesärztekammer:* Qualitätssicherung der quantitativen Bestimmungen im Laboratorium. Neue Richtlinien der Bundesärztekammer. Dt. Ärztebl. 85, A699–A712 (1988)

[17] *Bundesärztekammer:* Ergänzung der „Richtlinien der Bundesärztekammer zur Qualitätssicherung in medizinischen Laboratorien". Dt. Ärztebl. 91, C159–C161 (1994)
[18] *J. Angerer, T. Göen* and *G. Lehnert:* Mindestanforderungen an die Qualität von umweltmedizinisch-toxikologischen Analysen. Umweltmed. Forsch. Prax. 3, 307–312 (1998)
[19] *G. Lehnert, J. Angerer* and *K.H. Schaller:* Statusbericht über die externe Qualitätssicherung arbeits- und umweltmedizinisch-toxikologischer Analysen in biologischen Materialien. Arbeitsmed. Sozialmed. Umweltmed. 33(1), 21–26 (1998)
[20] *J. Angerer* and *G. Lehnert:* Anforderungen an arbeitsmedizinisch-toxikologische Analysen – Stand der Technik. Dt. Ärztebl. 37, C1753–C1760 (1997)

Author: *M. Ball*
Examiner: *J. Wittsiepe*

Addendum: Alternative extraction procedure

The deep-frozen blood sample is thawed in the water bath at 40 °C and homogenised by ultrasound (15 min) and shaking. Then 50 to 60 mL whole blood are measured in a measuring cylinder, weighed and transferred to a 500 mL bottle with a screw cap. After addition of 50 mL extracted bidistilled water and the solution of the internal standard, the contents are thoroughly mixed for 30 min on a tumbler shaker. At the same time a blank sample is prepared with 50 mL water and 20 µL dodecane.

Then 75 mL saturated ammonium sulphate solution are added and the sample is shaken intensively for 1 minute by hand. After addition of 75 mL ethanol, the sample is again shaken intensively for 1 minute. Then 100 mL n-hexane are added and the mixture is again shaken intensively for 3 minutes by hand.

The mixture is subsequently centrifuged for 5 minutes to accelerate separation of the phases. The n-hexane phase is withdrawn and passed through a drying column (length = 140 mm, diameter = 30 mm, filled with 20 g anhydrous sodium sulphate) into a 250 mL pear-shaped distilling flask that has been previously weighed on an analytical balance.

Once again 100 mL n-hexane are added to the aqueous phase and the extraction step is repeated. Then the drying column is rinsed with 20 mL n-hexane and the combined extracts are evaporated until dryness on the rotary evaporator at 40 °C and 280 mbar. The sample is further heated at a pressure of 20 mbar until a constant weight is attained.

The weight of the fat is determined by finally weighing the flask on the analytical balance. Then the instructions in Section 3.3 must be followed.

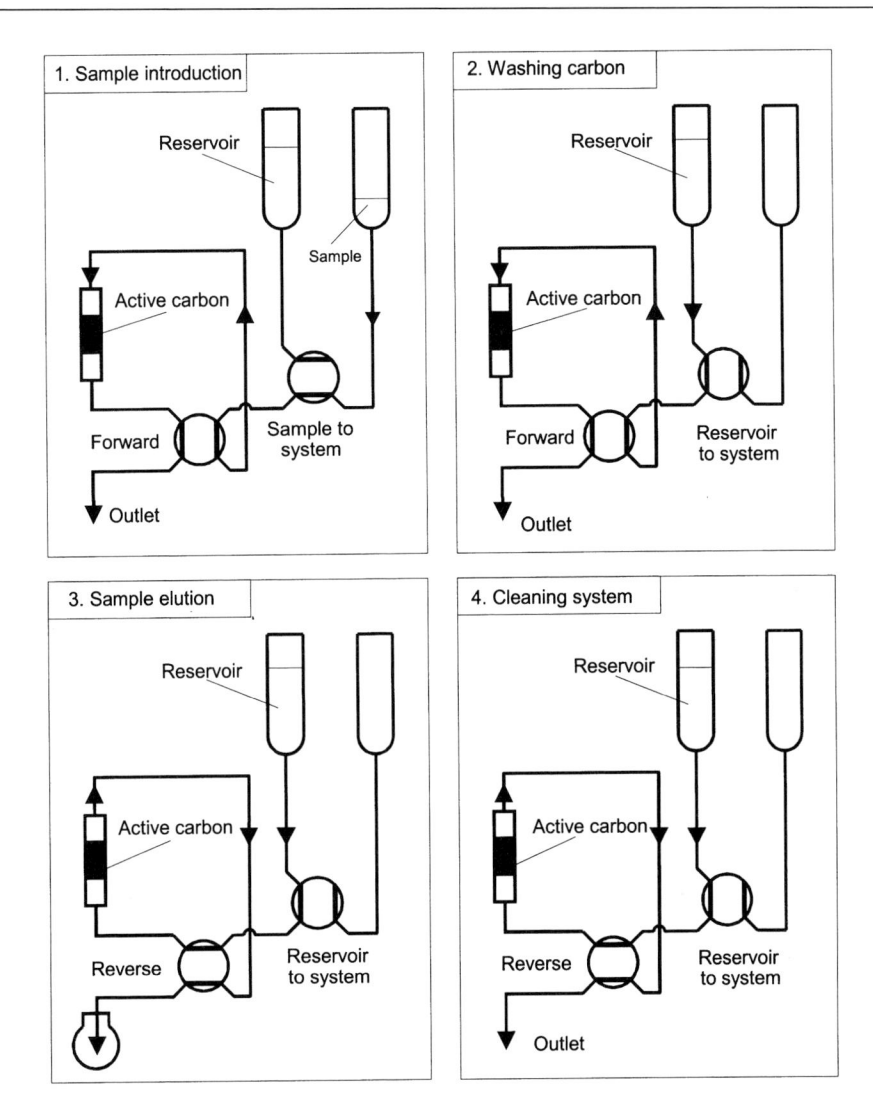

Fig. 1. Function of the active carbon microcolumn system from Ergo, Hamburg, Germany

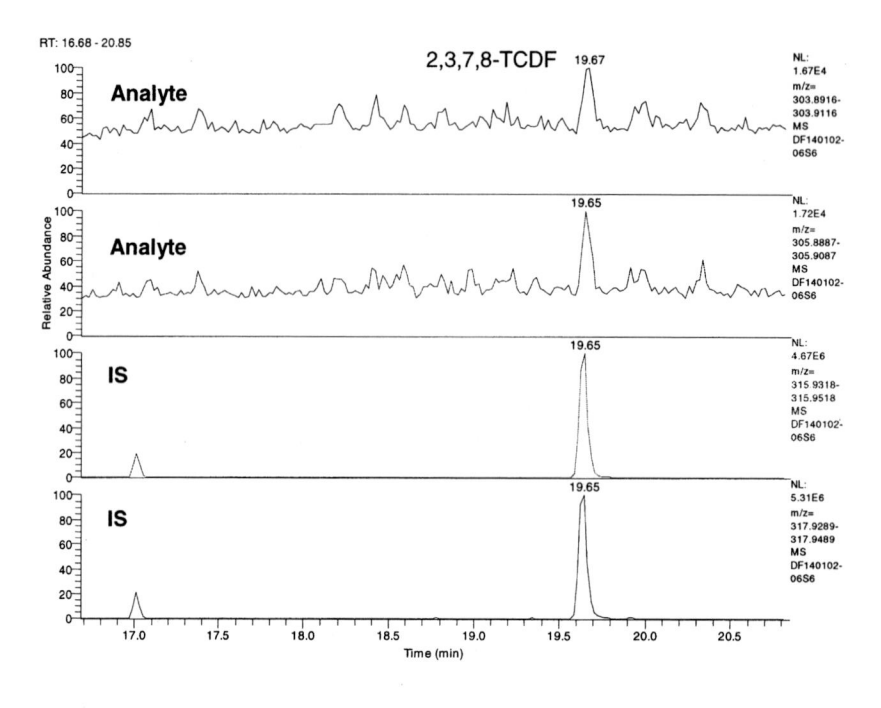

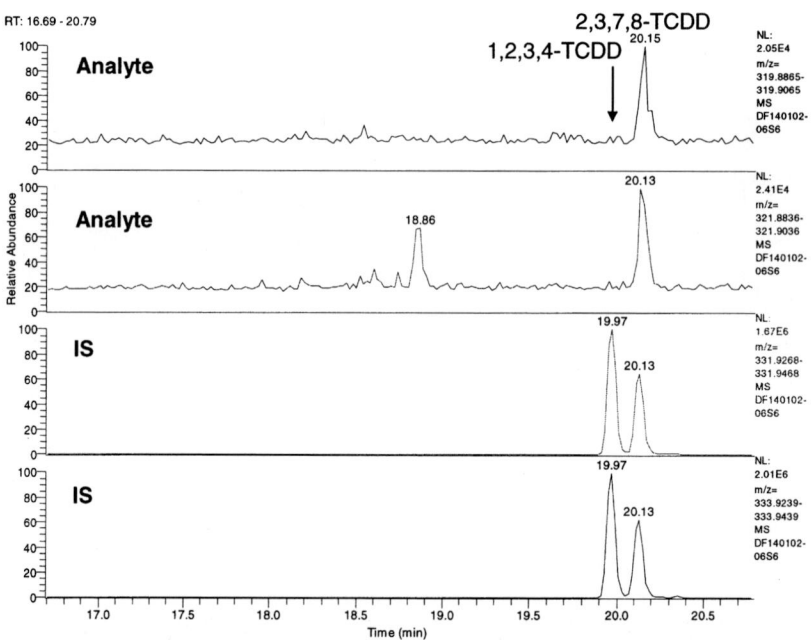

Fig. 2. Chromatogram of a processed native individual blood sample with a content of 43 pg WHO TEQ/g blood fat (Pe=penta, Hx=hexa, Hp=hepta, O=octa)

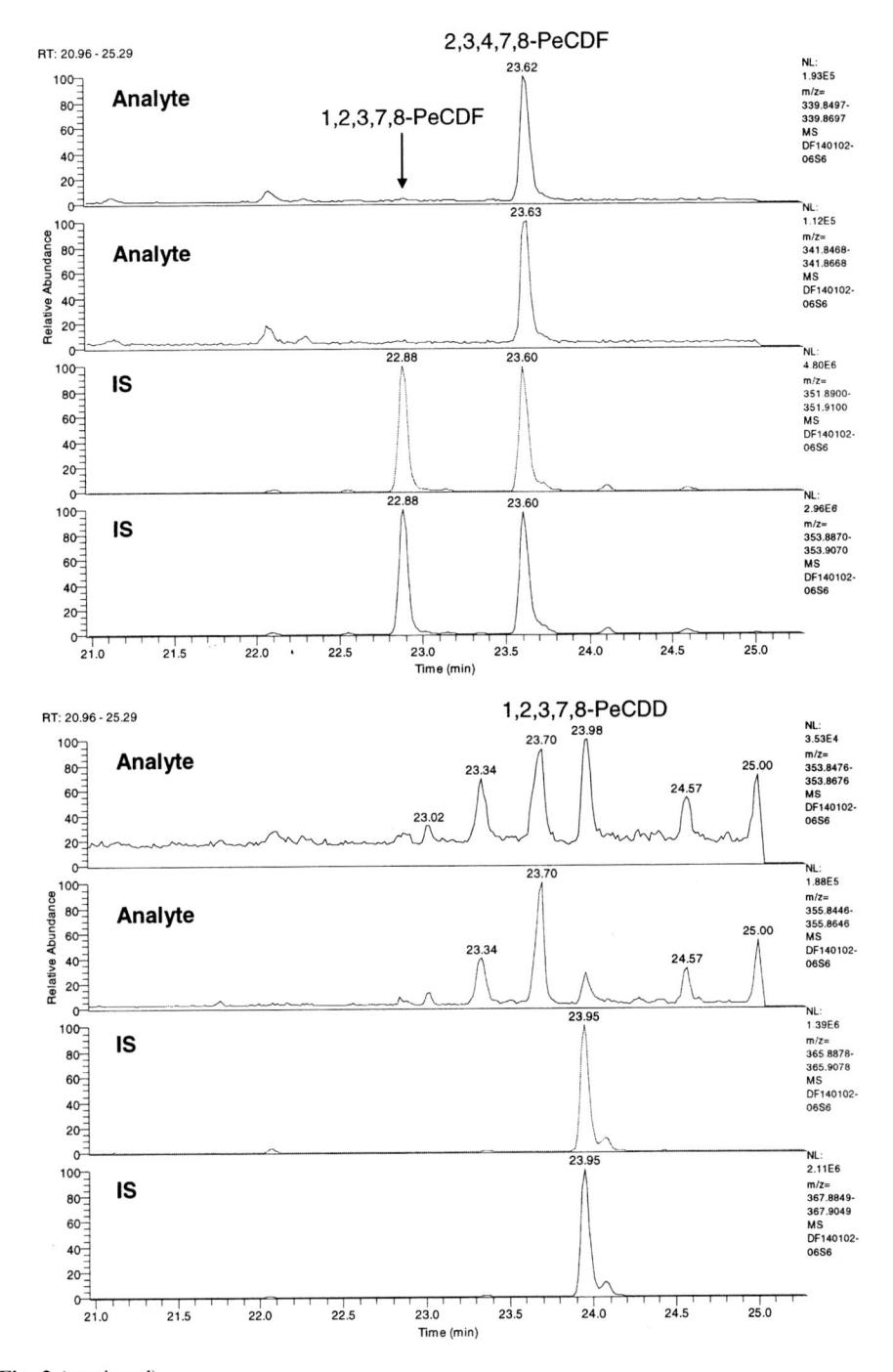

Fig. 2 (continued)

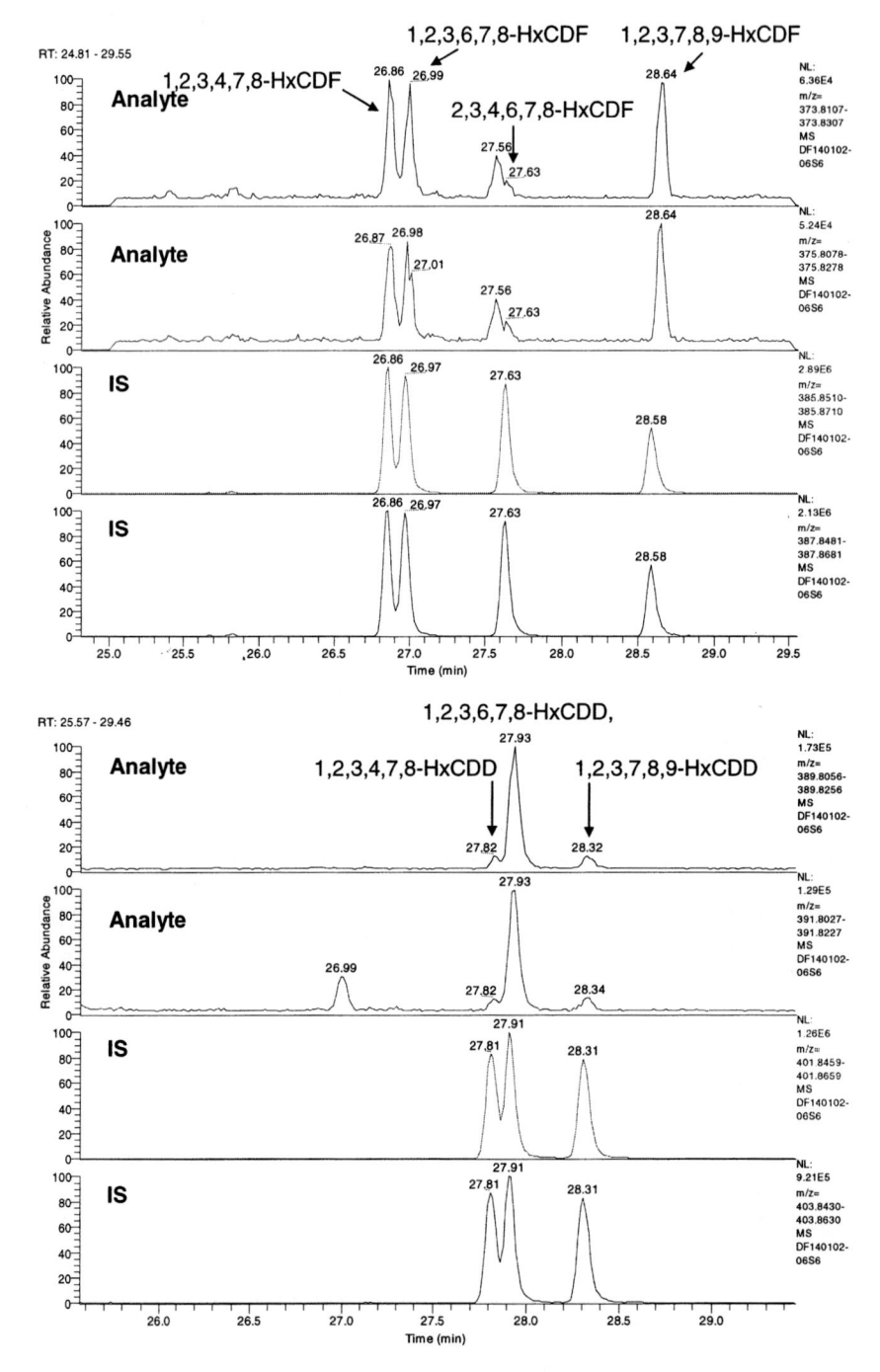

Fig. 2 (continued)

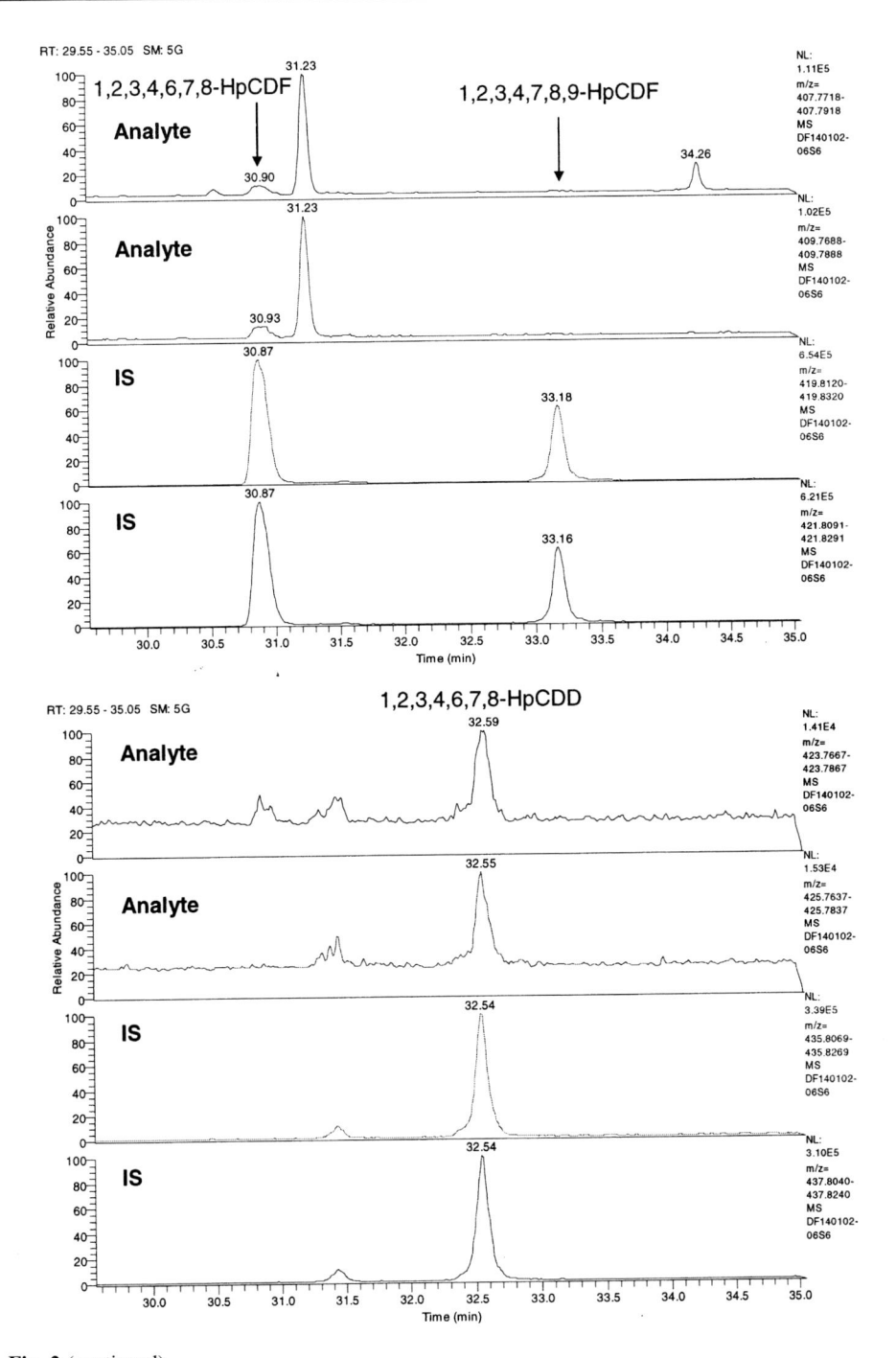

Fig. 2 (continued)

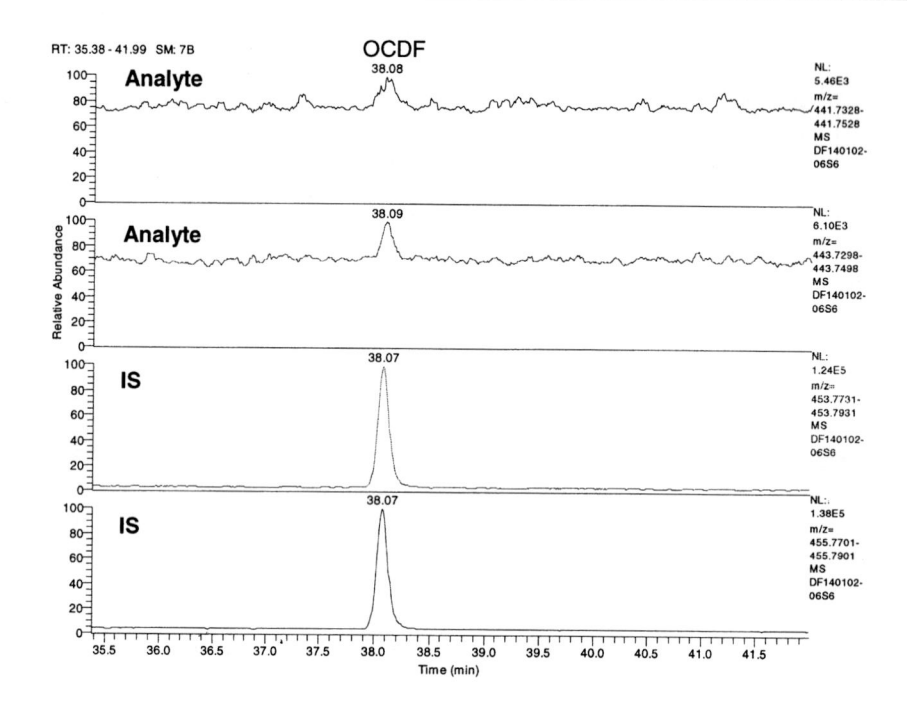

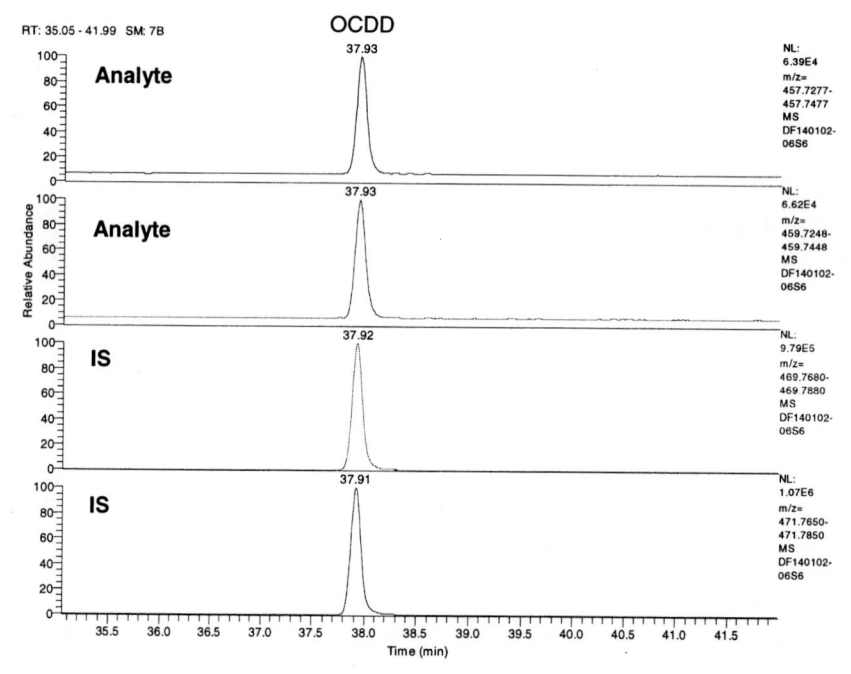

Fig. 2 (continued)

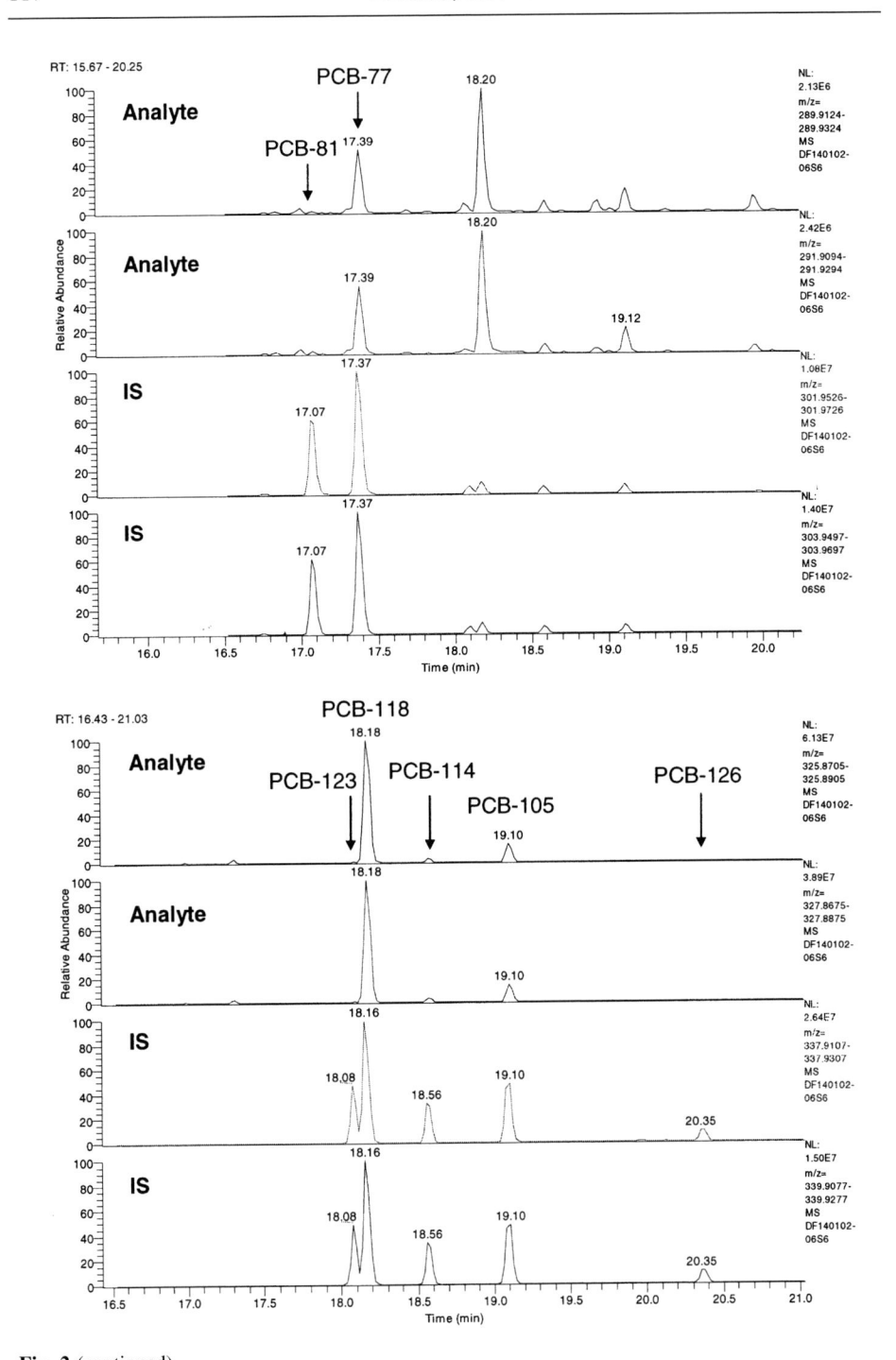

Fig. 2 (continued)

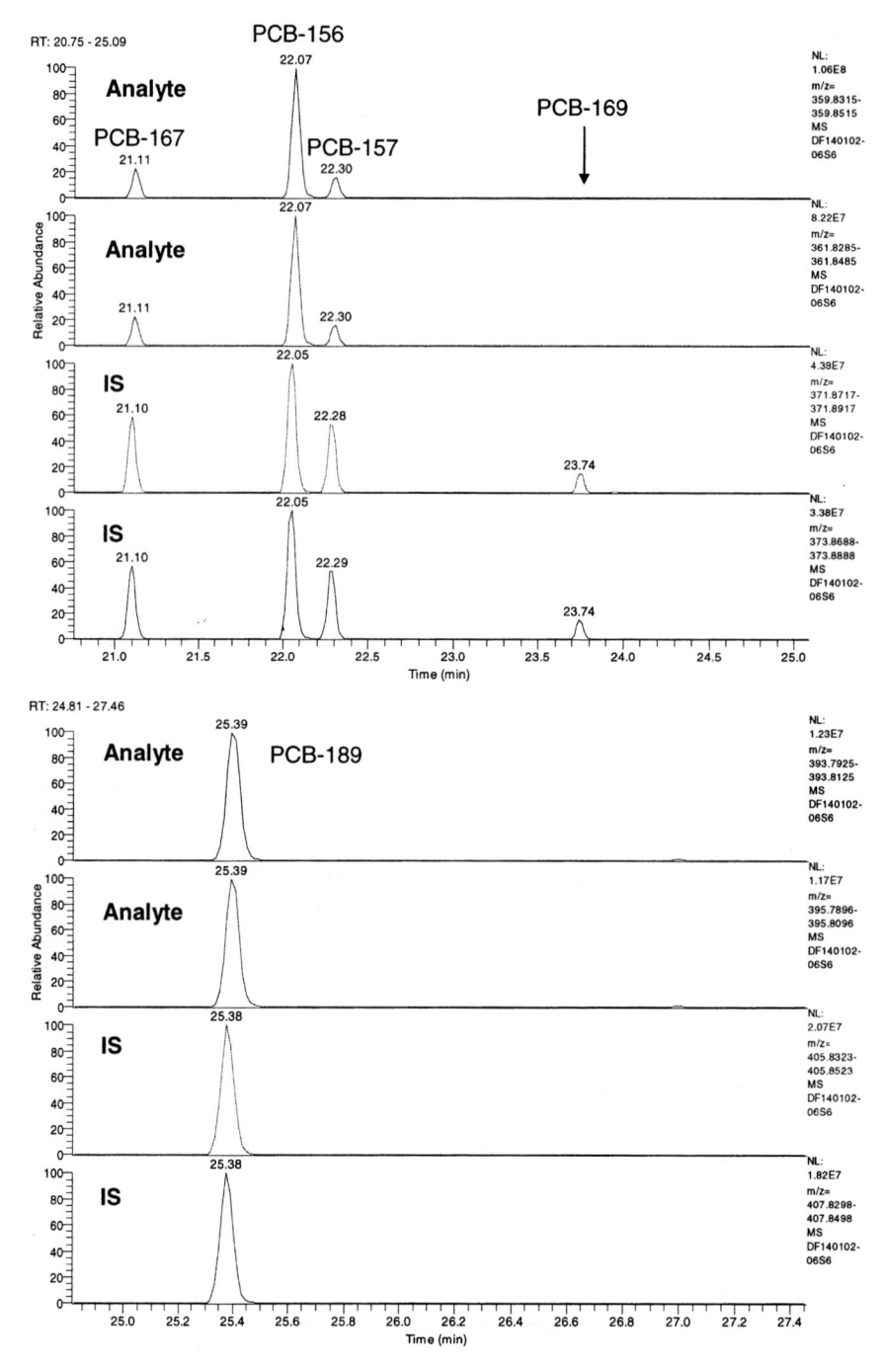

Fig. 2 (continued)

Hexamethylene diisocyanate (HDI) and hexamethylenediamine (HDA)

Application   Determination in urine

Analytical principle   Capillary gas chromatography/mass spectrometry

Completed in   March 2002

The method described here is suitable for the sensitive determination of hexamethylenediamine (HDA) in urine after occupational exposure to hexamethylene diisocyanate (HDI) or hexamethylenediamine.

After an acidic hydrolysis, the urine samples are adjusted to an alkaline pH value, and chloroformic acid ethyl ester is added. This converts the hexamethylenediamine conjugates formed from the absorbed HDA and by the metabolisation of HDI to hexamethylenediamine, which is then converted to the 1,6-bis-(ethoxycarbonylamino)-hexane (HDA-diurethane) derivative in a Schotten-Baumann reaction. The analyte derivatives are subsequently extracted in cyclohexane. The non-conjugated HDA that can be present in urine samples, especially after massive HDA exposure, can be derivatised in the same manner by omitting the acidic hydrolysis step. The generated hexamethylene diurethane is separated by means of capillary gas chromatography and determined by mass selective detection in the SIM mode. 1,7-Heptamethylenediamine serves as an internal standard. Calibration is carried out using calibration standards prepared in urine and treated in the same manner as the samples to be tested.

Hexamethylenediamine (HDA)

Within-series imprecision: Standard deviation (rel.) s_w = 3.5% or 1.8%
                               Prognostic range u = 7.8% or 4.0%
                               at a concentration of 10 µg or 100 µg HDA per litre urine
                               and where n = 10 determinations

Between-day imprecision: Standard deviation (rel.) s_w = 9.4% or 3.9%
                               Prognostic range u = 20.9% or 8.7%
                               at a concentration of 5 µg or 50 µg HDA per litre urine
                               and where n = 10 determinations

Accuracy:	Recovery rate	$r = 105\%$ or 98%
	at a concentration of 10 or 100 µg HDA per litre urine	
Detection limit:	2 µg HDA per litre urine	

Hexamethylene diisocyanate (HDI) and hexamethylenediamine (HDA)

OCN~~~~~NCO

1,6-Hexamethylene diisocyanate (HDI)

H₂N~~~~~NH₂

1,6-Hexamethylenediamine (HDA)

At room temperature 1,6-hexamethylene diisocyanate (CAS No. 822-06-0, synonyms: Lupranat H 201, hexane diisocyanate, diisocyanatohexane, hexamethylene-1,6-diisocyanate, hexamethylene diisocyanate, Desmodur H, Cardate) is a clear, colourless or slightly yellowish liquid with a boiling point of 255 °C. HDI is readily volatile, combustible, and has a low viscosity and a pungent odour. Hexamethylene diisocyanate is an important starting chemical in the manufacture of polyisocyanates, mainly the HDI dimer biuret and the HDI trimer isocyanurate. These pre-polymers are used as hardeners in dual-component polyurethane paints that are mainly used in the automobile construction industry. Much more than 90% of the HDI produced is used for this purpose. At the time of production the biuret pre-polymer contains about 0.7% of the HDI monomer. The monomer content can increase to approx. 1.6% during storage on account of *in situ* decomposition. The monomer content in the HDI trimer is approx. 0.2%, and as a rule it does not change during storage [1].

The worldwide use of HDI amounted to about 42,000–44,000 tonnes in 1991 [2].

The general population can occasionally be exposed to HDI by inhalation when using polyurethane paints. At the workplace there is a danger of intake via inhalation or through the skin during the production or application of the relevant polyurethane products, whereby exposure while applying polyurethane paints poses the most significant risk [2].

HDI is hydrolysed to HDA before, during and after absorption, and HDA undergoes a further biotransformation. However, these products were not detected in serum after experimental exposure of volunteers to 0.025 mg/m³ inhaled over a period of 7.5 hours [3] or to 0.005 mg/m³ over a period of 2 hours [4] (detection limit <0.1 µg/L), but they were excreted in urine at an elimination half-life of 1.1–1.4 hours [3] and 2.5 hours respectively [4]. In these studies an average of 11–21% [3] or 39% [4] of the administered dose was recovered as HDA conjugates in urine after HDI intake by inhalation. N-acetylation of the diamine was identified as the principle conjugation pathway, non-conjugated HDA was not detectable in the urine after intake by inhalation [4]. In another study, in which 6 volunteers were each administered an oral dose of 8.2 mg HDI, a mean value of 0.28 mg (1–6% of the dose) was found as

HDA in urine after hydrolysis [5]. There is at least one additional metabolic pathway in the case of oral intake. Oxidation of HDI by diaminoxidases leads not only to the formation of hexamethylenediamine but also of 6-aminohexanoic acid, which is also acetylated and excreted [5, 6]. On average 0.8 mg of the previously administered oral dose (8.2 mg) of this substance was recovered from the urine of the 6 test subjects [5, 7, 8].

From the point of view of occupational medicine, the effect of hexamethylene diisocyanate on the respiratory tract is of primary importance. Irritation to the eyes and respiratory tract can be caused by HDI. High concentrations lead to a decrease in the respiratory frequency, dyspnoea and respiratory insufficiency. This must be differentiated from non-specific or allergic bronchial or alveolar over-sensitivity. A skin-sensitising effect has been detected in animal studies. However, cases of allergic contact dermatitis have seldom been observed in occupational medicine. After repeated inhalation, and depending on the concentration and the exposure period, damage to the upper and lower respiratory tract has been caused in rats. This is manifested in the nasal cavity as hyperplasia/metaplasia and ulceration, while epithelisation, interstitial pneumonia or accumulation of macrophages in the alveoli is observed in the lungs. HDI is not mutagenic in the salmonella mutagenicity test with or without metabolic activation. No evidence of a carcinogenic effect of the substance was shown by long-term inhalation experiments on rats. Possible effects on reproduction due to exposure to HDI have not yet been investigated [9].

Only a few publications deal with the exposure of workers to HDI at the workplace. Thus the HDA concentration was determined in the plasma and urine of workers who handled adhesives containing HDI. No HDA was found in plasma or in urine (detection limit = 0.5 µg/L) at presumed HDI concentrations of <0.1 to 0.7 µg/m^3 in the ambient air (the air analyses were carried out several years previously) [10]. A further study examined 22 workers who applied paints containing HDI to automobile bodywork parts. The workers wore personal protective equipment during this activity. HDA was detected in the urine of 5 of the 22 persons at concentrations of between 1 and 12 µg HDA/g creatinine [11].

The Deutsche Forschungsgemeinschaft's Commission for the Investigation of Health Hazards of Chemical Compounds in the Work Area has stipulated a MAK value of 0.005 mL/m^3 (5 ppb) or 0.035 mg/m^3. The potential for sensitisation of the respiratory tract and the skin was indicated by the code *Sah* (danger of sensitisation of the respiratory tract and the skin) in the MAK list [9, 12].

Authors: *J. Lewalter, G. Skarping, D. Ellrich, U. Schoen*
Examiner: *G. Müller*

Hexamethylene diisocyanate (HDI) and hexamethylenediamine (HDA)

Application Determination in urine

Analytical principle Capillary gas chromatography/mass spectrometry

Completed in March 2002

Contents

1 General principles
2 Equipment, chemicals and solutions
2.1 Equipment
2.2 Chemicals
2.3 Solutions
2.4 Internal Standard (I.S.)
2.5 Calibration standards
3 Specimen collection and sample preparation
3.1 Sample preparation
4 Operational parameters for gas chromatography
5 Analytical determination
6 Calibration
7 Calculation of the analytical result
8 Standardisation and quality control
9 Evaluation of the method
9.1 Precision
9.2 Accuracy
9.3 Detection limit
9.4 Sources of error
10 Discussion of the method
11 References

1 General principles

After an acidic hydrolysis, the urine samples are adjusted to an alkaline pH value and chloroformic acid ethyl ester is added. This converts the hexamethylenediamine conjugates formed from the absorbed HDA and by metabolisation of HDI to hexamethylenediamine, which is then converted to the 1,6-bis-(ethoxycarbonylamino)-hexane (HDA-diurethane) derivative in a Schotten-Baumann reaction. The analyte derivatives are subsequently extracted in cyclohexane. The non-conjugated HDA that can be present in urine samples, especially after massive HDA exposure, can be derivatised in the same manner by omitting the acidic hydrolysis step. The generated hexamethylene diurethane is separated by means of capillary gas chromatography and determined by mass selective detection in the SIM mode. 1,7-Heptamethylenediamine serves as an internal standard. Calibration is carried out using calibration standards prepared in urine and treated in the same manner as the samples to be tested.

2 Equipment, chemicals and solutions

2.1 Equipment

Capillary gas chromatograph with mass-selective detector (MSD) and autosampler as well as an integrator or a PC system for data evaluation.
Separation column: DB-5MS (e.g. from Kupfer 128-5522); length: 25 m, inner diameter: 0.2 mm, film thickness: 0.33 µm

Heating block (dry thermostat) from Techne – DB-BB

pH meter

Vortex shaker (e.g. from Cenco, Netherlands)

Turbovap LV evaporator (from Zymark)

Dispensers adjustable from 0 to 5 mL

10 mL test-tubes (e.g. from Schütt Labortechnik – 356 1103) with screw caps and Teflon seals (e.g. from Schott)

100 mL Volumetric flasks

5 mL Centrifuge tubes (e.g. from Sarstedt) 73.705

200 µL Conical microvials (e.g. from Macherey-Nagel: 70286)

3.5 mL Transfer pipettes (e.g. from Sarstedt)

Microlitre pipettes, adjustable between 1 and 10 µL, 10 and 100 µL, and 100 and 1000 µL (e.g. from Eppendorf).

2.2 Chemicals

1,6-Hexamethylenediamine p. a. (e. g. from Aldrich; H1.169-6)

1,7-Heptamethylenediamine p. a. (e. g. from Sigma; D3266)

Chloroformic acid ethyl ester, p. a. (e. g. from Merck 800881)

37% Hydrochloric acid p. a. (e. g. from Merck 1.00317.1000)

25% Ammonia solution p. a. (e. g. from Fluka; 09860)

32% Sodium hydroxide p. a. (e. g. from Merck 1.05587.2500)

Cyclohexane p. a. (e. g. from Merck; 1.09666.1000)

Helium 99.999%

Ultrapure water or bidistilled water

Ethanol, p. a. (e. g. from Merck)

Toluene for trace analysis (e. g. from Merck).

2.3 Solutions

2.4 Internal Standard (I.S.)

Stock solution:
Approximately 10 mg heptamethylenediamine are weighed exactly into a 100 mL volumetric flask. The volumetric flask is then filled to its nominal volume with ethanol (100 mg/L).

Spiking solution I.S.:
10 mL of the stock solution of the internal standard are pipetted into a 100 mL volumetric flask. The flask is subsequently filled to its nominal volume with water (10 mg/L).

2.5 Calibration standards

Stock solution:
10 mg hexamethylenediamine are weighed exactly into a 100 mL volumetric flask. The volumetric flask is then filled to its nominal volume with ethanol while swirling the contents around occasionally. This solution can be stored in the refrigerator for at least 2 weeks (100 mg/L).

Working solution:
10 mL of the stock solution are pipetted into a 100 mL volumetric flask. The flask is subsequently filled to its nominal volume with water (10 mg/L).

Calibration standards:
Calibration standards in pooled urine are prepared from the stock solution and the working solution in accordance with the following pipetting scheme. For this purpose the volumes of the solutions shown in Table 1 are each filled into a 50 mL volumetric flask, and then the flask is filled to its nominal volume with pooled urine. This material is divided into 2 mL aliquots which are pipetted into 5 mL sealable polyethylene tubes and stored at approx. −18 °C. Calibration standards thus stored are stable for at least 2 months.

Table 1. Pipetting scheme for the preparation of calibration standards in pooled urine in 50 mL volumetric flasks

Volume of the stock solution [µL]	Volume of the working solution [µL]	Concentration of the calibration standard [µg/L]
–	–	0
–	25	5
–	100	20
–	250	50
100	–	200

3 Specimen collection and sample preparation

The specimens should be collected at the end of a working shift. The urine samples are collected in plastic bottles. If it is not possible to process the samples immediately, the urine can be stored for several months at approx. −18 °C in the deep-freezer (tested for periods up to 3 months) until processing can be carried out.

3.1 Sample preparation

Before analysis, the samples are thawed (if necessary) and thoroughly mixed. 1 mL urine is pipetted into a 10 mL test-tube with a screw cap and a Teflon seal, and 10 µL of the spiking solution of the internal standard are added. 0.5 ml 37% hydrochloric acid is added using a pipette, and the samples are incubated for 2 hours at 100 °C in a water bath. After cooling, the samples are adjusted to a pH value of >12 by careful addition of 150 µL 25% ammonia solution and 1.4 mL 32% sodium hydroxide. The pH value is checked using a pH meter. 2.0 mL cyclohexane and 100 µL chloroformic acid ethyl ester are added, and the samples are shaken for 1 minute intensively by means of the Vortex mixer. Then the samples are allowed to stand for 1 minute. The process of shaking and allowing to stand is carried out a total of 3 times. In each case the tube must be shaken immediately after addition of the derivatisation agent and the cyclohexane. Then the sample is centrifuged for 10 minutes at 3000 g to separate the phases. The upper cyclohexane phase is withdrawn using a pipette and transferred to an evaporation tube (Teflon). The sample in the tube is evaporated to dryness in the vacuum centrifuge. The residue is then taken up in 50 µL toluene, and transferred to a microvial for the subsequent analysis.

4 Operational parameters for gas chromatography

Capillary column:	Material:	Fused silica
	Stationary phase:	DB 5 MS
	Length:	25 m
	Inner diameter:	0.2 mm
	Film thickness:	0.33 µm
Detector:	Mass selective detector (MSD)	
Temperatures:	Column:	Starting temperature 100 °C isothermal for 2 minutes, then increase at a rate of 10 °C/min to 280 °C, 5 minutes at the final temperature
	Injector:	280 °C
	Detector:	300 °C
Carrier gas:	Helium 5.0 at a flow rate of 3.7 mL/min	
Split:	Splitless, split on after 1 minute at 35 mL/min	
Septum purging:	3 mL/min for 5 minutes	
Injection volume:	1 µL	
Ionisation type:	Electron impact ionisation (EI)	
Ionisation energy:	70 eV	
Dwell time:	30 ms	
Electron multiplier:	1800–2000 V	

All other parameters must be optimised in accordance with the manufacturer's instructions.

The retention times given below were found for the analyte and the internal standard under the chromatographic conditions described here. They serve only as a guide. Users of the method must satisfy themselves of the separation power of the GC column they use and of the resulting retention behaviour of the substances.

Figure 1 shows the chromatogram of a native urine sample containing 42 µg HDA/L.

5 Analytical determination

In order to carry out the gas chromatographic analysis, 1 µL of each of the prepared urine samples is injected into the GC. If the measured values are outside the range of the calibration graph, the samples are diluted accordingly and processed anew.
A quality control sample is analysed with each analytical series.
Quantification of the individual aromatic amines is based on the masses shown in Table 2.

Table 2. Retention times and masses used for quantification

Compound	Retention time [minutes]	Mass [m/z]
Hexamethylenediamine (as 1,6-bis-(ethoxycarbonylamino)-hexane)	17.19	158/215/260*
Heptamethylenediamine (as 1,6-bis-(ethoxycarbonylamino)-heptane)	18.20	229/274*

The masses marked with * are used for quantitative evaluation.

6 Calibration

The calibration standards are processed in the same manner as the urine samples (Section 3.1) and analysed by means of gas chromatography/mass spectrometry as described in Sections 4 and 5. Linear calibration graphs are obtained by plotting the quotients of the peak areas of analyte and that of the internal standard as a function of the concentrations used. It is unnecessary to plot a complete calibration graph for every analytical series. It is sufficient to include one calibration standard in the analysis per day. The ratio of the results obtained for this standard and the result for the equivalent standard in the complete calibration graph is calculated. Using this quotient, each result read off the calibration graph is adjusted for the relevant series.
New calibration graphs should be plotted if the quality control results indicate systematic deviation.
The calibration graph is linear between the detection limit and 250 µg per litre urine. Figure 2 shows a calibration function of HDA in urine.

7 Calculation of the analytical result

Quotients are calculated by dividing the peak areas of the analyte by that of the internal standard. These quotients are used to read off the concentration of HDA in µg per litre from the relevant calibration graph.

8 Standardisation and quality control

Quality control of the analytical results is carried out as stipulated in the guidelines of the Bundesärztekammer (German Medical Association) [13, 14] and in the special preliminary remarks to this series. In order to determine the precision of the method a urine sample containing a constant concentration of hexamethylenediamine is analysed. As material for quality control is not commercially available, it must be prepared in the laboratory. For this purpose, pooled urine is spiked with a defined quantity of HDA. The concentration of this control material should lie within the relevant concentration range. A six-month supply of the control material is prepared, divided into aliquots in 10 mL polyethylene tubes and stored in the deep-freezer. The theoretical value and the tolerance range for this quality control material are determined in the course of a pre-analytical period (one analysis of the control material on each of 20 different days) [15–17].

9 Evaluation of the method

9.1 Precision

In order to determine the precision in the series pooled urine of test persons who had not been exposed to HDA or HDI was spiked with 10 µg/L or 100 µg/L and then processed and analysed. Ten replicate determinations of the urine samples yielded the precision in the series shown in Table 3.

Table 3. Precision in the series for the determination of HDA (n = 10)

Substance	Concentration [µg/L]	Standard deviation (rel.) [%]	Prognostic range [%]
HDA	10	3.5	7.8
	100	1.8	4.0

In addition, the precision from day to day was determined. The same material was used as for the determination of the precision in the series. These urine samples were processed and analysed on each of 10 different days. The precision from day to day is given in Table 4.

Table 4. Precision from day to day for the determination of HDA (n = 10)

Substance	Concentration [µg/L]	Standard deviation (rel.) [%]	Prognostic range [%]
HDA	5	9.4	20.9
	50	3.9	8.7

9.2 Accuracy

The accuracy of the method was checked by recovery experiments and by round-robin experiments. The same material that was used to determine the precision in the series was analysed to measure the recovery rate, and this material was processed and analysed 10 times as described in the working instructions. The results were evaluated using calibration standard solutions prepared with another pool of urine. The relative recovery rate was 105% at a concentration of 10 µg/L and 98% at a concentration of 100 µg/L.

9.3 Detection limit

The detection limit was 2 µg HDA per litre urine under the given conditions for sample processing and chromatographic determination at a signal/background ratio of at least 3:1.

9.4 Sources of error

Due to the selected gas chromatographic separation of the diamine derivative and in the case of GC-MSD evaluation of one specific ion fragment no interference was observed. If, despite expectations to the contrary, interference occurs, the qualitative and quantitative evaluation can be ensured by taking several mass fragments of the HDA derivative into account.

10 Discussion of the method

The method presented here is based on an analytical method published by G. Skarping et al. in 1989 [6]. The method permits biomonitoring of the aliphatic diamine in urine samples from people who handle hexamethylenediamine (HDA) directly as well as those who handle hexamethylene diisocyanate (HDI), which is subsequently metabolised to HDA. Direct extraction of the aliphatic diamine leads to high losses, but this can be avoided by direct derivatisation of the free HDA present in urine or of HDA released from its conjugates by prior hydrolysis, and then by subsequent extraction of the derivative. Despite changing the sample processing from the conditions given for this method in the literature [6], comparable results were achieved. Less solvent is used for sample preparation in this case. This permits a multi-shaker to be used. Thus it is possible to process many samples simultaneously. As the diurethane is lipophilic, it was possible to successfully use cyclohexane instead of toluene, whereby the time required for evaporation of the samples was considerably shortened. No decomposition of the HDA diurethane was observed during chromatography, even at high temperatures. The present method is practicable and permits up to 80 urine samples containing the readily soluble HDA to be reproducibly and highly sensitively analysed per day. This method uses commonly used laboratory equipment. If a halogenated derivatising agent is used, the evaluation of the analysis can also be carried out on a GC-ECD.

The extraction of the diamines in the form of the diurethanes permits a valid HDA analysis down to the trace level. Other derivatisation procedures using substances such as PFPA (pentafluoropropionic anhydride) etc. are unsuitable, as they require primary extraction of the hydrophilic HDA, which leads to high losses.

Instruments used:
Agilent 5890 gas chromatograph with Agilent 5970B mass selective detector,
Agilent 7673A autosampler, Hewlett-Packard Vectra XM Series 4 data station 5/133 and Agilent MSD Chemstation evaluation software.

11 References

[1] *P.M. Hulse:* An evaluation of HDI in polyurethane spray paint aerosols. NTIS No. AD-A151. 606. (1984)

[2] *Gesellschaft Deutscher Chemiker / Beratergremium für umweltrelevante Altstoffe (BUA):* Hexamethylendiisocyanat. BUA-Stoffbericht 112. Verlag S. Hirzel, Stuttgart (1993)

[3] *T. Brorson, G. Skarping* and *J. Nielsen:* Biological monitoring of isocyanates and related amines. II. Test chamber exposure of humans to 1,6-hexamethylene diisocyanate. Int. Arch. Occup. Environ. Health 62, 385–389 (1990)

[4] *H. Tinnerberg, G. Skarping, M. Dalene* and *L. Hagmar:* Test chamber exposure of humans to 1,6-hexamethylene diisocyanate and isophorone diisocyanate. Int. Arch. Occup. Environ Health 67(6), 367–374 (1995)

[5] *T. Brorson, G. Skarping, J.F. Sandstrom* and *M. Stenberg:* Biological monitoring of isocyanates and related amines. I. Determination of 1,6-hexamethylene diamine (HDA) in hydrolysed human urine after oral administration of HDA. Int. Arch. Occup. Environ. Health 62(1), 79–84 (1990)

[6] *G. Skarping, M. Dalene, T. Brorson, J.F. Sandström, C. Sangö* and *A. Tiljander:* Chromatographic determination of amines in biological fluids with special reference to the biological monitoring of isocyanates and amines. J. Chromatogr. 479 125–133 (1989)

[7] *G. Skarping, M. Dalene, B.-G. Svensson, M. Littorin, B. Akesson, H. Welinder* and *S. Skerfving:* Biomarkers of exposure, antibodies, and respiratory symptoms in workers heating polyurethane glue. Occup. Environ. Med. 53, 180–187 (1996)

[8] *M. Au, W.F. Diller, M. Heger, H.-D. Hoffmann, R. Rühl, B. Scheel* and *V. Wilms:* Sicherer Umgang mit isocyanathaltigen Produkten – Vorschläge zur Erfassung der Exposition und Verbesserung der Prävention. Zbl. Arbeitsmed. 50, 335–341 (2000)

[9] *H. Greim (ed.):* Hexamethylendiisocyanate. Arbeitsmedizinisch-toxikologische Begründung von MAK- und BAT-Werten, 23rd ed. Wiley-VCH, Weinheim (1996)

[10] *M. Littorin, L. Rylander, G. Skarping, M. Dalene, H. Welinder, U. Stromberg* and *S. Skerfving:* Exposure biomarkers and risk from gluing and heating of polyurethane: a cross-sectional study of respiratory symptoms. Occup. Environ. Med. 57(6), 396–405 (2000)

[11] *N.R. Williams, K. Jones* and *J. Cocker:* Biological monitoring to assess exposure from use of isocyanates in motor vehicle repair. Occup. Environ. Med. 56(9), 598–601 (1999)

[12] *Deutsche Forschungsgemeinschaft:* MAK- und BAT-Werte-Liste, Report 37. Wiley-VCH, Weinheim (2001)

[13] *Bundesärztekammer:* Qualitätssicherung der quantitativen Bestimmungen im Laboratorium. Neue Richtlinien der Bundesärztekammer. Dt. Ärztebl. 85, A699–A712 (1988)

[14] *Bundesärztekammer:* Ergänzung der „Richtlinien der Bundesärztekammer zur Qualitätssicherung in medizinischen Laboratorien". Dt. Ärztebl. 91, C159–C161 (1994)

[15] *G. Lehnert, J. Angerer* and *K.H. Schaller:* Statusbericht über die externe Qualitätssicherung arbeits- und umweltmedizinisch-toxikologischer Analysen in biologischen Materialien. Arbeitsmed. Sozialmed. Umweltmed. 33(1), 21–26 (1998)

[16] *J. Angerer* and *G. Lehnert:* Anforderungen an arbeitsmedizinisch-toxikologische Analysen – Stand der Technik. Dt. Ärztebl. 37, C1753–C1760 (1997)

[17] *J. Angerer, T. Göen* and *G. Lehnert:* Mindestanforderungen an die Qualität von umweltmedizinisch-toxikologischen Analysen. Umweltmed. Forsch. Prax. 3, 307–312 (1998)

Authors: *J. Lewalter, G. Skarping, D. Ellrich, U. Schoen*
Examiner: *G. Müller*

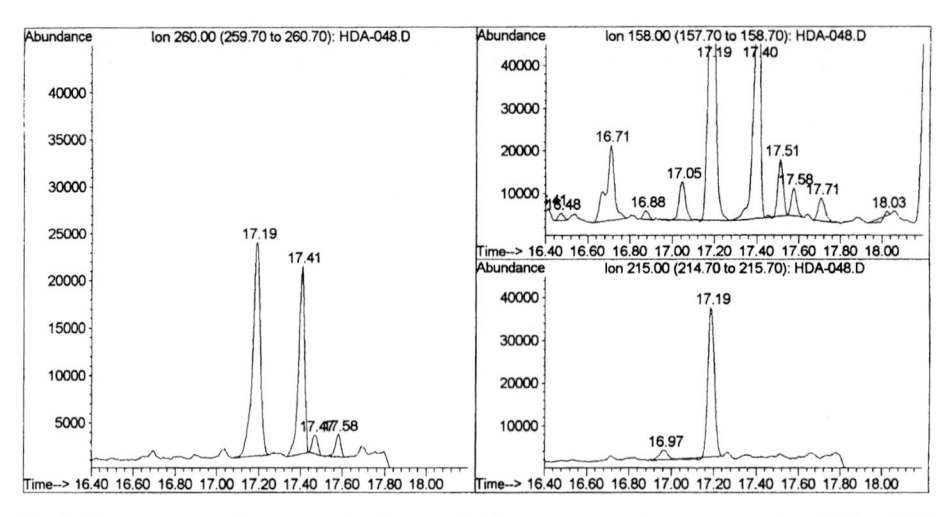

Hexamethylene diisocyanate (HDI) and hexamethylenediamine (HDA)

Fig. 1. Chromatogram of a processed urine sample from a person who was exposed to HDI or HDA at the workplace (c = 42 µg/L)

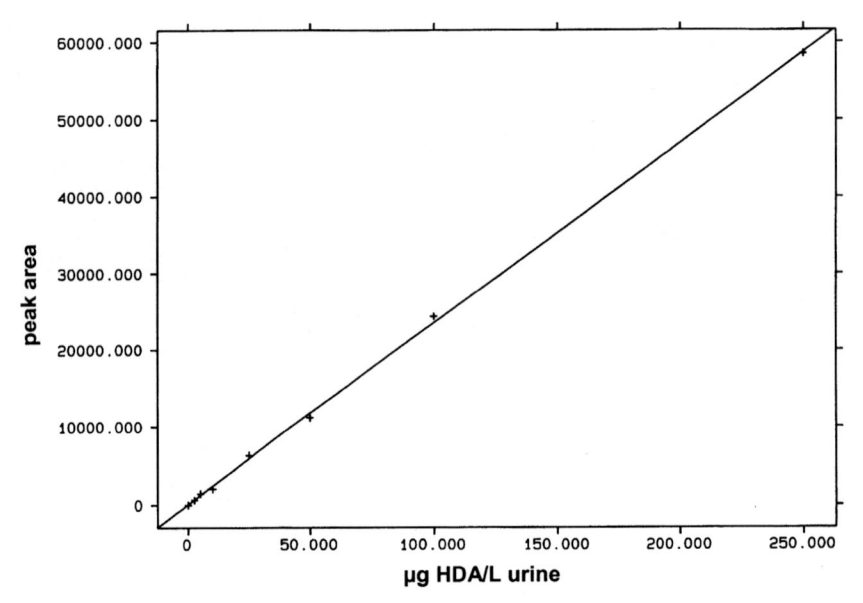

Fig. 2. Calibration function of HDA in urine

8-Hydroxy-2′-deoxyguanosine

Application   Determination in urine

Analytical principle   High performance liquid chromatography/ electrochemical detection

Completed in   January 2002

Summary

This method serves to quantify 8-hydroxy-2′-deoxyguanosine, which is formed from the reaction of various reactive oxygen species with human genetic material and is excreted in the urine after DNA repair.

8-Hydroxy-2′-deoxyguanosine (8-OHdG) is separated from acidified urine by liquid/solid chromatography in sample preparation columns filled with phenyl-modified silica gel. After elution of the solid phase, 8-OHdG is separated by high performance liquid chromatography and quantified by means of an electrochemical detector. Calibration is performed using calibration standards that are prepared in pooled urine and are treated in the same manner as the samples to be analysed.

8-Hydroxy-2′-deoxyguanosine (8-OHdG)

Within-series imprecision: Standard deviation (rel.) s_w = 8.7% or 9.5%
Prognostic range u = 19.4% or 21.2%
at a concentration of 10 µg or 100 µg 8-OHdG per litre urine
and where n = 10 determinations

Between-day imprecision: Standard deviation (rel.) s_w = 9.3% or 11.2%
Prognostic range u = 20.6% or 24.8%
at a concentration of 10 µg or 100 µg 8-OHdG per litre urine
and where n = 10 determinations

Accuracy: Recovery rate r = 108% at 10 µg/L and 94% at 100 µg/L

Detection limit: 5 µg 8-OHdG per litre urine

8-Hydroxy-2′-deoxyguanosine (8-OHdG)

Reactive oxygen species (ROS), such as singlet oxygen, ozone, hydrogen peroxide, superoxide radical anions and hydroxyl radicals, are formed by different *endogenous* processes (e.g. by incomplete reduction of oxygen in the respiratory chain) and *exogenous* processes (e.g. due to ionising radiation, inorganic and organic toxic substances and their metabolites).

If ROS react with the genetic material, some of the bases in the DNA strands may be oxidised. Covalent DNA adducts are formed.

8-Hydroxy-2′-deoxyguanosine [CAS number: 88847-89-6] (8-OHdG, Synonyms: 8-oxo-7,8-dihydro-2′-deoxyguanosine, 8-oxo-dG) is such a covalent modified nucleoside of the genetic material. Of the base adducts that are formed by reactions with ROS it is the oxidation product of DNA that occurs in the largest quantities.

Guanine modified in this manner then pairs with adenine instead of cytosine, and this leads to G:C → T:A transversions in bacterial and mammalian cells. However, ROS can also cause oxidation in the nucleotide pool of the cells. The deoxyribonucleoside-5′-triphosphate of 8-OHdG generated there is paired with adenine during DNA synthesis and after repair and renewed replication A:T → C:G transversions are caused [1].

Such DNA adducts may lead to mutations and cause the initiation of cancer cells if they are not repaired. Therefore they represent a marker for the initiation and possibly also for the progression in the multiple-step concept of carcinogenesis [2].

As already mentioned, ROS may also be formed endogenously during cell respiration. Therefore the organism has developed protective mechanisms to prevent damage to DNA or to repair it. ROS are intercepted by various enzymatic (superoxide dismutase, catalase, glutathione peroxidase) and non-enzymatic (vitamin E, glutathione, ascorbate) antioxidants. In general, oxidative stress arises when the equilibrium between the oxidants and the endogenous antioxidants shifts in favour of the oxidants.

Moreover, cells exhibit a series of repair mechanisms to protect themselves against the genotoxic effects of 8-OHdG on the DNA strand. For instance *Escherichia coli* has three repair enzymes:

1) Formamidopyridine-DNA-glycosylase (Fpg also known as MutM, a glycosylase and AP-lyase) excises 8-hydroxyguanine (8-OHGua) when it is paired with cytosine [3, 4].
2) MutY (a monofunctional glycosylase) excises adenine from the 8-OHGua:adenine erroneous pairing [5].
3) MutT (a triphosphatase) hydrolyses 8-OHGua-triphosphate to the harmless monophosphate [1].

Fpg and MutY function on the principle of base excision repair (BER). During BER erroneous or abnormal DNA bases are excised from the DNA in the form of free modified bases by DNA glycosylases. Apuric or apyrimidic (AP) sites remain in the DNA. AP endonucleases subsequently remove the deoxyribose phosphate groups, and individual nucleotide gaps remain. These gaps are then filled again by DNA polymerases [6, 7]. Recently homologues of such enzymes have also been discovered in mammalian cells [8–12].

There is an additional single-step repair mechanism in humans (*nucleotide excision repair*) in which the entire modified nucleotide (e.g. 8-OHdG) is removed from the DNA [13].

Both 8-OHdG and 8-hydroxyguanine (8-OHGua) are repair products of the oxidised guanine base from DNA. 8-OHGua is the main excretion product of DNA repair in both the eukaryotes and the prokaryotes. At least 90% are excised in the form of 8-OHGua by the glycosylases [14].

Although 8-OHGua is excreted in higher concentrations than 8-OHdG, it is not suitable as a biomarker for oxidative stress because its excretion concentration is strongly influenced by individual nutritional habits. Rats, for example, excrete about 85% less 8-OHGua when fed on a diet without nucleic acids than when they have a "normal" diet. In contrast, the renal excretion of 8-OHdG is not significantly affected by diet in rats [15, 16]. In addition, as 8-OHdG is scarcely metabolised further in the organism and it is renally excreted, its determination is a suitable diagnostic tool for biological monitoring.

The mean quantity of 8-OHdG excreted by humans is approximately 60 ng/kg body weight or 1.5 to 4.5 µg/L urine per day [17]. Other publications report normal excretion in the range of between 2 and 15 µg per litre urine [18].

Sixty-nine test subjects (41 men, 28 women) of different ages were investigated at intervals of 4 to 8 weeks in six independent measurement series in 1996 [19] (Table 1). The main objective of the study was to establish the baseline value and to investigate the intraindividual and interindividual differences in 8-OHdG excretion. In addition, the influence of verifiable and measurable confounders such as age, gender, body mass index, smoking and passive smoking on 8-OHdG excretion was investigated.

Table 1. Investigation results on 8-OHdG excretion in urine

8-OHdG [µg/g creatinine]	n	Investigated groups of persons	Reference
2.52±0.95	21	Non-exposed non-smokers	Tagesson et al. 1993 [18]
5.25±1.88	5	Non-smokers exposed to diazo-dyes	
3.45±1.45	9	Non-smokers exposed to asbestos	
3.98±1.32	16	Workers in the rubber industry, non-smokers	
2.82±1.12	20	Non-exposed smokers	
4.71±2.2	25	Smokers exposed to diazo-dyes	
3.53±1.40	21	Non-smokers exposed to asbestos	
3.3±1.20	12	Workers in the rubber industry, non-smokers	
3.40±1.22	65	Smokers and non-smokers exposed to benzene	Lagorio et al. 1994 [17]
28.41±17.40	14	SCC (small-cell carcinoma) lung cancer patients	Erhola et al. 1997 [21]
19.77±17.40	23	Non-SCC lung cancer patients	
19.38±18.49	52	Control group	
6.75±4.71	60	Healthy non-smokers	Germadnik et al. 1997 [19]
35.80±16.40	81	Diabetics not dependent on insulin	Leinonen et al. 1997 [23]
24.28±15.19	100	Control group	
10.8 (5.5–12.3)	27	Strong chromate reducers (5.4 (<1.0–7.3) µg Cr from erythrocytes/L blood)	Lewalter 1999 [24]
32.5 (10.2–71.3)	54	Weak chromate reducers (13.8 (5.2–26.8) µg Cr from erythrocytes/L blood)	
9.2 (<5.0–10.6)	50	Control group (1.1 (<1.0–2.0))	
5.88±4.23	85	Children with Down's syndrome	Jovanovic et al. 1998 [25]
3.37±2.6	81	Children in the control group (living in the same household)	
		Mean values in 24 h urine from 6 measurement series	Germadnik et al. 1997 [19]
4.85±1.00	23	Smokers	
3.9±1.03	22	Passive smokers	
3.96±1.08	21	Non-smokers	
4.16±1.13	40	Men	
4.4±1.08	26	Women	
4.26±1.10	67	Total	
3.93		Medians of 8-OHdG in 24 h urine for investigations 1 to 6	
3.30		Investigation 1	
2.93		Investigation 2	
4.33		Investigation 3	
4.11		Investigation 4	
4.03		Investigation 5	
4.28		Investigation 6	
		All investigations	

The mean values calculated from the 6 measurements showed that on average smokers excreted 25% more 8-OHdG than non-smokers or passive smokers. In contrast, no significant difference could be found between the baseline excretion of passive smokers and non-smokers.

Highly significant differences in the concentrations of 8-OHdG were determined between the individual examination dates. Based on the six independent measurement series an average interindividual VC of 48% (18 to 107%) and an average intraindividual VC of 36% in a range of 7.4 to 44.9 nmol/24 h were calculated for 8-OHdG excretion. Men excrete 27% more 8-OHdG in urine than women. However, these differences disappear when the values are based on the individual body weights. If the values were standardised on the creatinine excretion, then women excreted slightly more 8-OHdG than men.

The confounders smoking, gender, weight, size and age have a distinct influence on the 8-OHdG excretion per 24 h. However, if the values are based on the body weight, only smoking and age remain as dominant confounders.

The vitamin levels of each test person were also measured (vitamins A, C and E as well as β-carotene). Highly significant negative correlations for β-carotene and vitamin E with 8-OHdG excretion were found. No relationship between the vitamin A and C levels and 8-OHdG excretion was observed [19].

The determination of 8-OHdG in urine is very suitable for biological monitoring of groups of people who have been exposed over a long period to substances that are known to be mutagenic or carcinogenic and that are assumed to act by means of an ROS-related mechanism. These groups (e.g. workers in the azo-dye, rubber and asbestos industries [20], the chromate industry [24], or patients suffering from cancer who have recently undergone chemotherapy or radiotherapy [21]) exhibit considerably elevated concentrations of 8-OHdG in the urine compared with unexposed control groups (Table 1). Great fluctuations in 8-OHdG excretion were observed within a group. This can be attributed to the individual differences in tobacco consumption and age, and also to gender [19], diet [22], illnesses (e.g. diabetes mellitus [23]) and the genetic characteristics of the test persons.

Authors: *J. Lewalter, G. Leng, M. Wimber*
Examiners: *G. Scherer, T. Renner*

8-Hydroxy-2′-deoxyguanosine

Application   Determination in urine

Analytical principle   High performance liquid chromatography/ electrochemical detection

Completed in   January 2002

Contents

1   General principles
2   Equipment, chemicals and solutions
2.1 Equipment
2.2 Chemicals
2.3 Solutions and conditioning
2.4 Calibration standards
3   Specimen collection and sample preparation
3.1 Sample preparation
4   Operational parameters
4.1 Operational parameters for high performance liquid chromatography
5   Analytical determination
6   Calibration
7   Calculation of the analytical result
8   Standardisation and quality control
9   Evaluation of the method
9.1 Precision
9.2 Accuracy
9.3 Detection limits
9.4 Sources of error
10 Discussion of the method
11 References

1 General principles

8-Hydroxy-2′-deoxyguanosine (8-OHdG) is separated from acidified urine by liquid/ solid chromatography in sample preparation columns filled with phenyl-modified silica gel. After elution of the solid phase, 8-OHdG is separated by high performance liquid chromatography and quantified by means of an electrochemical detector. Calibration is performed using calibration standards that are prepared in pooled urine and are treated in the same manner as the samples to be analysed.

2 Equipment, chemicals and solutions

2.1 Equipment

HPLC system consisting of a gradient pump, a device for degassing the eluents, a column thermostat, an injection valve, an autosampler, an electrochemical detector (working electrode: graphite, reference electrode: Ag/AgCl, e.g. HP-ECD 1049 A from Agilent) and an integrator or a PC system for data evaluation.

High performance liquid chromatographic column:
Zorbax Bonus RP, length: 250 mm, inner diameter: 4.6 mm; particle diameter: 5 µm (e.g. Agilent Zorbax Bonus RP Order No.: 880668-901)

Test-tube shaker (Vortex mixer) (e.g. from IKA)

Adjustable pipette 10–100 µL (e.g. from Eppendorf)

Adjustable pipette 100–1000 µL (e.g. from Eppendorf)

Adjustable pipette 1000–5000 µL (e.g. from Eppendorf)

10 mL Polyethylene tubes

Sealable plastic bottles for sample storage

Volumetric flasks 20, 25, 100, 2000 mL

Laboratory centrifuge (e.g. from Hettich)

Microvials for the HPLC autosampler

2.2 Chemicals

Phenyl cartridge 3 mL, 500 mg (e.g. from Baker, Art. No. 7095-03, Bakerbond spe phenyl C_6H_5)

8-Hydroxy-2'-deoxyguanosine (e.g. from Sigma, H-5653)

Formic acid for trace analysis (e.g. Suprapur from Merck)

Acetonitrile (e.g. Lichrosolv from Merck)

Na_2SO_4 anhydrous p.a. (e.g. from Riedel-de Haën)

Methanol (e.g. from Merck)

Perchloric acid, p.a. (e.g. from Merck)

Deionised water (e.g. produced by means of Millipore® technology)

2.3 Solutions and conditioning

Eluent A (40% acetonitrile and 60% H_2O):
1200 mL water are placed in a 2 L volumetric flask. The volumetric flask is filled to its nominal volume with acetonitrile while swirling the contents at regular intervals.

Eluent B (0.1% formic acid):
Add 2 mL concentrated formic acid to 2000 mL water.

0.4 M Perchloric acid:
Approx. 50 mL water are placed in a 100 mL volumetric flask. 3.4 mL concentrated perchloric acid are added using a pipette, and the flask is filled to its nominal volume with water.

Conditioning of the phenyl cartridge:
The 3 mL phenyl cartridges (500 mg) are conditioned by introducing 2.5 mL methanol and then 2.5 mL water and allowing the solvents to run slowly through the cartridges. The cartridges are subsequently dried by suction.

Elution solution for the phenyl cartridges (pH 2):
Approx. 70 mL water are placed in a glass beaker. Formic acid is carefully added using a pipette while monitoring the pH by means of a pH meter and stirring until a pH value of 2 is reached. Then 40 mL methanol are added to 60 mL of this formic acid solution in a glass beaker, and the contents are thoroughly mixed.

2.4 Calibration standards

The calibration standards are prepared in pooled urine. Pooled urine is prepared from spontaneous urine samples collected in a suitable vessel. The samples are thoroughly mixed and stored at $-18\,°C$ until the standards and the control material are prepared.

Stock solution:
Approximately 10 mg 8-OHdG are weighed exactly into a 100 mL volumetric flask. The flask is subsequently filled to its nominal volume with water (100 mg/L).

Working solution A:
10 mL of the stock solution are pipetted into a 100 mL volumetric flask. The flask is subsequently filled to its nominal volume with water (10 mg/L).

Working solution B:
1.0 mL of the stock solution is pipetted into a 100 mL volumetric flask. The flask is subsequently filled to its nominal volume with water (1 mg/L).

Calibration standards:
Calibration standards in pooled urine are prepared in accordance with the following pipetting scheme. For this purpose the volumes of the relevant working solutions shown in Table 2 are each filled into a 50 mL volumetric flask, and then the flask is filled to its nominal volume with pooled urine. This material is divided into 6 mL

aliquots which are pipetted into 10 mL sealable polyethylene tubes and stored at approx. −18 °C. Calibration standards thus stored are stable for at least 2 months.

Table 2. Pipetting scheme for the preparation of calibration standards in pooled urine in 50 mL volumetric flasks

Volume of working solution A [μL]	Volume of working solution B [μL]	Concentration of the calibration standard [μg/L]
–	–	0
–	50	1*
–	150	3*
–	250	5
–	500	10
100	–	20
250	–	50
500	–	100

*) The detection limit is dependent on the quantity of urine used (see Section 9.3).

3 Specimen collection and sample preparation

The urine is collected in sealable plastic bottles. The urine samples should be processed as soon as possible, at the latest after 3 days of storage at 4 °C. If the samples cannot be processed immediately, the urine must be stored in the deep-freezer at approx. −20 °C until processing can be carried out. The stability of frozen urine samples was checked over a period of 2 months.

3.1 Sample preparation

Before analysis, the samples are thawed (if necessary) and thoroughly mixed. Then 5 mL urine are pipetted into a 20 mL test-tube, and the pH is adjusted to a value of 5 by adding 0.4 M $HClO_4$ (the pH value is monitored by means of a pH electrode; if the pH value is too low, 0.1 M NaOH must be added). The solution is subsequently transferred to a pre-treated phenyl cartridge. It is eluted using 9 mL water. The aqueous eluate is collected and transferred to a second *fresh* phenyl cartridge that has been conditioned as described above. Any eluate which drips out during this operation is discarded. Then the cartridge is eluted with 1 mL of a mixture of 60 parts of water at pH=2 and 40 parts of methanol into an autosampler vial for the subsequent HPLC analysis.

4 Operational parameters

4.1 Operational parameters for high performance liquid chromatography

Separation column:	Material:	Steel
	Length:	250 mm
	Inner diameter:	4.6 mm
	Column packing:	Zorbax Bonus RP, 5 µm

Separation principle: Reversed phase

Temperature: 40 °C

Detection: Electrochemical detector (amperometry)
Graphite working electrode
Reference electrode with internal electrolytes (Ag/AgCl)
Potential 700 mV

Mobile phase: Eluent A: 40% CH_3CN and 60% H_2O
Eluent B: 0.1% formic acid
(The eluents must be degassed before use!)

Gradient: 11 min 96% B, 4% A
12 min 10% B, 90% A
15 min 10% B, 90% A
20 min 96% B, 4% A
25 min 96% B, 4% A

Flow rate: 0.8 mL/min

Injection volume: 50 µL

Retention time of 8-OHdG: 9 minutes

All other parameters must be optimised in accordance with the manufacturer's instructions.

A retention time of 9.0 minutes was found for 8-hydroxy-2'-deoxyguanosine under the chromatographic conditions described here. This retention time serves only as a guideline. Users of the method must satisfy themselves of the separation power of the HPLC column they use and of the resulting retention behaviour of the substance. Figure 1 shows the HPLC chromatograms of a native urine pool and several calibration standards in urine.

5 Analytical determination

In each case 50 µL of the urine samples processed as described in Section 3 are injected into the HPLC instrument. If the measured values are not within the linear range of the calibration curve, the urine samples are diluted with ultrapure water and processed anew.
A quality control sample is analysed with each analytical series.

6 Calibration

The calibration standards are processed in the same manner as the urine samples (Section 3.1) and analysed by means of high performance liquid chromatography with nitrogen-specific detection as described in Sections 4 and 5. In each case 50 µL of the processed calibration standards in urine are injected into the HPLC.
The calibration curve is obtained by plotting the integrated peak area of 8-OHdG as a function of the concentrations used. It is unnecessary to plot a complete calibration graph for every analytical series. It is sufficient to analyse one calibration standard for every analytical series. The ratio of the result obtained for this standard and the result for the equivalent standard in the complete calibration graph is calculated. Using this quotient, each result read off the calibration graph is corrected.
New calibration graphs should be plotted if systematic deviations are observed in the results of the precision check and if a new mobile phase or a new electrode on the detector is used.
Urine samples with an 8-OHdG concentration of between 5 µg/L and 50 µg/L can be measured without interference under the conditions described here and using this detector.

7 Calculation of the analytical result

The resulting peak areas of the analyte are used to read off the corresponding concentration of 8-OHdG in µg per litre urine from the relevant calibration curve. As a rule, the pooled urine used to prepare the calibration standards exhibits a background signal, and so the resulting calibration graph has to be shifted in parallel so that it passes through the zero point of the coordinates. The concentration of 8-OHdG in the urine used to prepare the calibration standards can be read off the intercept with the axis before the parallel shift is carried out.

8 Standardisation and quality control

Quality control of the analytical results is carried out as stipulated in the guidelines of the Bundesärztekammer (German Medical Association) [26, 27] and in the special preliminary remarks to this series. In order to check the precision of the method a urine control sample containing a constant concentration of 8-OHdG is analysed. As material for quality control is not commercially available, it must be prepared in the laboratory. It is advisable to use urine or pooled urine with a native concentration of 8-OHdG of approx. 10 µg/L, or to spike the urine of non-exposed persons with 10 µg/L. A six-month supply of this control material is prepared, divided into aliquots in 10 mL polyethylene tubes and stored in the deep-freezer. The theoretical value and the tolerance range for this quality control material are determined in the course of a pre-analytical period (one analysis of the control material on each of 20 different days) [28–30].

9 Evaluation of the method

9.1 Precision

Pooled urine spiked with 10 µg/L and 100 µg/L of 8-OHdG was processed and analysed as described in the previous sections to check the precision in the series. Ten replicate determinations of the urine samples yielded the precision in the series documented in Table 3.

In addition, the precision from day to day was determined. The same material was used as for the determination of the precision in the series. This urine was processed and analysed on each of 10 different days. The precision results are also shown in Table 3.

Table 3. Precision for the determination of 8-OHdG

	n	Concentration [µg/L]	Standard deviation (rel.) [%]	Prognostic range [%]
In the series	10	10	8.7	19.4
	10	100	9.5	21.2
From day to day	10	10	9.3	20.6
	10	100	11.2	24.8

9.2 Accuracy

Recovery experiments were carried out to test the accuracy of the method. Urine was spiked with 10 or 100 µg 8-OHdG per litre for this purpose. Each of these solutions was subsequently processed and analysed ten times in accordance with Section 3. The mean relative recovery rates were 108% (10 µg/L) and 94% (100 µg/L).

In addition, the losses due to processing were checked as a measure of the accuracy. For this purpose reference standards prepared in urine were processed and analysed. Aqueous standards with the same quantity of 8-OHdG were simultaneously prepared in the same manner as the calibration standards. These aqueous standards were injected into the HPLC and analysed without further treatment. The mean recovery rate of 86.6% (range 85–90%) at a concentration of 10 µg/L was found by comparison of the peak areas of the aqueous standards with the peak areas of the 10 processed calibration standards in urine. This means that an average of 13.4% of 8-OHdG is lost during processing of the urine samples.

9.3 Detection limits

Under the conditions given here the detection limit, calculated as three times the signal/noise ratio of the analytical background in the temporal vicinity of the analyte signal, is approximately 5 µg/L and is limited by the background noise. Many of the urine samples investigated by this method did not exhibit this background interference. In such cases it was possible to detect 8-OHdG concentrations as low as 1 µg/L with certainty.

9.4 Sources of error

Interfering peaks which impaired the quantification of 8-OHdG below a value of 5 µg/L were observed in the vicinity of the 8-OHdG peak in some of the analysed urine samples. 8-OHdG background levels are given as 2 to 15 µg/L urine or µg/g creatinine in the literature. During optimisation of the method it became apparent that the selected phenyl cartridge and the HPLC separation column had a decisive influence on the reproducibility, the separation of interfering peaks and the cleaning-up of the solutions for analysis. Therefore the phenyl cartridges should be checked by batch before use, and the quantity of elution solutions should be adjusted as required. As the method is carried out without an internal standard, it is essential to treat the samples in an identical manner.

It is absolutely necessary to ensure that the eluents are thoroughly degassed, preferably by means of helium or by using a degasser. In general, the baseline of the ECD must be readjusted after each separation run. Electronic compensation was carried out by the HPLC instrument used in this case. Apart from thorough cleansing of the graphite electrode no further cleaning steps are generally necessary. However, we expressly refer the reader to the appropriate instructions given by the manufacturer of the electrochemical detection cell.

10 Discussion of the method

As a rule, the analytical method presented here is capable of reliably and reproducibly determining 8-OHdG excreted in urine in the range of the lower background levels of 1 to 2 µg/L urine. However, interfering peaks were observed in some urine samples. Although the sample clean-up with phenyl cartridges largely eliminated the interference, this still limited the detection limit of the method to 5 µg/L.

The sample preparation by means of phenyl cartridges functions according to chromatographic principles. The analyte is separated from part of the urine matrix in the first phenyl cartridge. As the analyte is also slightly retained, the first column must be eluted with relatively large amounts of water (9 mL). The analyte is subsequently enriched on the second phenyl cartridge, and then it is eluted using only 1 mL of an acidic methanol/water mixture (pH 2). Thus the analyte can be enriched by a factor of 5. The quantities of elution solutions given here represent an optimum for the phenyl cartridges used by the author of the method. As the properties of the solid phase material can vary not only from one manufacturer to another but also between batches, the user of the method must satisfy himself of the separation power of the material used.

The examiner of the method established that the procedure is readily replicated and verified its reliability. The reliability criteria are excellent, especially in view of the fact that the method had to be carried out without using an internal standard. A suitable internal standard is not available at present. Up to 40 determinations of 8-OHdG can be carried out and evaluated per day by an experienced analyst.

As described, the sensitive and selective determination of 8-OHdG in urine was achieved using an HPLC with electrochemical detection (ECD). Compared with UV detection, this detection is more sensitive for 8-OHdG by a factor of approximately 1000 [31, 32], and moreover it is relatively inexpensive to purchase. However, before measurement is carried out using the electrochemical detector various cleaning steps are necessary due to the strong influence of the urine matrix (see also e.g. [17, 19, 20]). Recent research on the determination of 8-OHdG in urine also includes the use of the HPLC-MS technique (e.g. [33]). In this case clean-up of the sample may be unnecessary due to the use of column-combination techniques. The use of this type of detection is also possible with the sample processing described here, provided slight modifications are made. In this case 5 mL urine are passed through only one conditioned (see Section 2.3) phenyl cartridge, eluted with methanol/water (40:60, pH 2) and then measured. The mass transfer 283.9 → 167.85 in the positive electrospray mode is recorded. Deuterated 8-OHdG is recommended as the internal standard in the LC-MS/MS procedure. The interference which may sometimes occur in an ECD has not been observed when the LC-MS technique is used. Independent of the urine matrix, the detection limit of 1 µg 8-OHdG per litre urine required to evaluate the background level can be reliably achieved without interference.

Instruments used:
HPLC HP 1090 from Agilent with integrated autosampler, electrochemical detector HP-ECD 1049 A, evaluation software: HP Chemstation for LC Rev. A.0603[509].

11 References

[1] M. Inoue, H. Kamiya, K. Fujikawa, Y. Ootsuyama, N. Murata-Kamiya, T. Osaki, K. Yaumoto and H. Kasai: Induction of chromosomal gene mutations in Escherichia coli by direct incorporation of oxidatively damaged nucleotides – new evaluation method for mutagenesis by damaged DNA precursors in vivo. J. Biol. Chem. 273, 11069–11074 (1998)

[2] E. Eder: DNA-Addukte krebserregender Stoffe in der Umwelt und am Arbeitsplatz: Bildung, Nachweis und Bedeutung. Umweltmed. Forsch. Prax. 4, 323–334 (1999)

[3] S. Boiteux, T. R. O'Connor, F. Lederer, A. Gouyette and J. Laval: Homogeneous Escherichia coli FPG Protein – A DNA glycosylase which excises imidazole ring-opened purines and DNA at apuric/apyrimidic sites. J. Biol. Chem. 265, 3916–3922 (1990)

[4] P. T. Henderson, J. C. Delanney, F. Gu, S. T. Tannenbaum and J. M. Essigmann: Oxidation of 7,8-dihydro-8-oxoguanine affords lesions that are potent sources of replication in vivo. Biochemistry 41, 914–921 (2002)

[5] L. F. Agnez-Lima, R. L. Napoletano, R. P. P. Fuchs, P. Di Mascio, A. R. Muori and C. F. M. Menck: DNA repair and sequence context affect. Part 1: O_2-induced mutagenesis in bacteria. Nucl. Acid. Res. 29, 2899–2903 (2001)

[6] R. M. Elliott, S. B. Astley, S. Southon and D. B. Archer: Measurement of cellular repair activities for oxidative DNA damage. Free Radical Biol. Med. 28, 1438–1446 (2000)

[7] S. Loft and H. E. Poulsen: Cancer risk and oxidative damage in man. J. Mol. Med. 74, 297–312 (1996)

[8] T. K. Hazra, T. Izumi, L. Maidt, R. A. Floyd and S. Mitra: The presence of two distinct 8-oxoguanine repair enzymes in human cells: Their potential complementary roles in preventing mutations. Nucl. Acids Res. 26, 5116–5122 (1998)

[9] K. Bialkowski and K. S. Kasprzak: Activity of the antimutagenic enzyme 8-oxo-2'-deoxyguanosine 5'-triphosphate pyrophosphohydrolase (8-Oxo-dGTPase) in cultured Chinese hamster ovary cells: effects of cell cycle, proliferation rate, and population density. Free radical Biol. Med. 28, 337–344 (2000)

[10] T. A. Rosenquist, D. O. Zharkov and P. Grollman: Cloning and characterization of a mammalian 8-oxoguanine DNA glycosylase. Proc. Natl. Acad. Sci. 94, 7429–7434 (1997)

[11] M. Pflaum, O. Will and B. Epe: Determination of steady-state levels of oxidative DNA base modifications in mammalian cells by repair endonucleases. Carcinogenesis 18, 2225–2231 (1997)

[12] K. Shinmura, S. Yamaguchi, T. Saitho, M. Takeuchi-Sasaki, S.-R. Kim, T. Nohmi and J. Yokota: Adenine excisional repair function of protein on the adenine:8-hydroxyguanine base repair in double-stranded DNA. Nucl. Acid. Res. 28, 4912–4918 (2000)

[13] T. Bessho, K. Tano, H. Kasai, E. Otsuka and S. Nishimura: Evidence for two DNA repair enzymes for 8-hydroxyguanine (7,8-dihydro-8-oxo-guanine) in human cells. J. Biol. Chem. 268, 19416–19421 (1993)

[14] P. Degan, M. K. Shigenaga, E.-M. Park, P. E. Alperin and B. N. Ames: Immunoaffinity isolation of urinary 8-hydroxy-2'-deoxyguanosine and 8-hydroxyguanine and quantitation of 8-hydroxy-2'-deoxyguanosine in DNA by polyclonal antibodies. Proc. Natl. Acad. Sci. 89(8), 3375–3379 (1992)

[15] C. Fraga, M. K. Shigeneaga and B. N. Ames: Oxidative Damage to DNA during aging: 8-hydroxy-2'-deoxyguanosine in rat organs and urine. Proc. Natl. Acad. Sci. 87, 4533–4537 (1990)

[16] H. J. Helbock, K. B. Beckman, M. K. Shigenaga, P. B. Walter, A. A. Woodall, H. C. Yeo and B. N. Ames: DNA oxidation matters: The HPLC-electrochemical detection assay of 8-oxo-deoxyguanosine and 8-oxo-guanine. Proc. Natl. Acad. Sci. 95, 288–293 (1998)

[17] S. Lagorio, C. Tagesson, F. Forastiere, I. Iavarone, O. Axelson and A. Carere: Exposure to benzene and urinary concentrations of 8-hydroxydeoxyguanosine, a biological marker of oxidative damage to DNA. Occup. Environ. Med. 51, 739–743 (1994)

[18] C. Tagesson, M. Källberg and P. Leanderson: Determination of urinary 8-Hydroxydeoxyguanosine by coupled-column high-performance liquid chromatography with electrochemical detec-

tion: a non-invasive assay for in vivo oxidative DNA damage in humans. Toxicology Methods 1, 242–251 (1992)
[19] D. Germadnik: 8-Hydroxy-2'-deoxyguanosin im Harn als Biomarker für oxidativen Stress: Evaluierung einer Längsschnittstudie. Dissertation, Universität Wien, Austria (1999)
[20] C. Tagesson, D. Chabiuk, O. Axelson, B. Baranski, J. Palus and K. Wyszynska: Increased urinary excretion of the oxidative DNA adduct, 8-hydroxydeoxyguanosine, as a possible early indicator of occupational cancer hazards in the asbestos, rubber, and azo-dye industries. Pol. J. Occup. Med. Environ. Health 6, 357–368 (1993)
[21] M. Erhola, S. Toyokuni, K. Okada, T. Tanaka, H. Hiai, H. Ochi, K. Uccida, T. Osawa, M. N. Nieminen, H. Alho and P. Kellokumpu-Lehtinen: Biomarker evidence of DNA oxidation in lung cancer patients: association of urinary 8-hydroxy-2'-deoxyguanosine excretion with radiotherapy, chemotherapy, and response treatment. FEBS Letters 409, 287–291 (1997)
[22] B. M. Lee, S. K. Lee and H. S. Kim: Inhibition of oxidative DNA damage, 8-OHdG, and carbonyl contents in smokers treated with antioxidants (vitamin E, vitamin C, β-carotene and red ginseng). Cancer Lett. 132, 219–227 (1998)
[23] J. Leinonen, T. Lethimäki, S. Toyokuni, K. Okaba, T. Tanaka, H. Hiai, H. Ochi, P. Laippala, V. Rantalaiho, O. Wirta, A. Pasternack and H. Alho: New biomarker evidence of oxidative DNA damage in patients with non-insulin-dependent diabetes mellitus. FEBS Letters 417, 150–152 (1997)
[24] J. Lewalter and G. Leng: Charakterisierung und Bewertung individueller Beanspruchungen beim industriellen Arbeitsstoffumgang (8-Hydroxy-desoxy-Guanosin (8-HOdG)-Ausscheidung nach Alkali-Chromat-Umgang). Reports of the 40th Annual Congress of the DGAUM, Berlin (2000)
[25] S. V. Jovanovic, D. Clements and K. MacLeod: Biomarkers of oxidative stress are significantly elevated in Down's syndrome. Free Rad. Biol. Med. 25, 1044–1048 (1998)
[26] Bundesärztekammer: Qualitätssicherung der quantitativen Bestimmungen im Laboratorium. Neue Richtlinien der Bundesärztekammer. Dt. Ärztebl. 85, A699–A712 (1988)
[27] Bundesärztekammer: Ergänzung der „Richtlinien der Bundesärztekammer zur Qualitätssicherung in medizinischen Laboratorien". Dt. Ärztebl. 91, C159–C161 (1994)
[28] G. Lehnert, J. Angerer and K. H. Schaller: Statusbericht über die externe Qualitätssicherung arbeits- und umweltmedizinisch-toxikologischer Analysen in biologischen Materialien. Arbeitsmed. Sozialmed. Umweltmed. 33(1), 21–26 (1998)
[29] J. Angerer and G. Lehnert: Anforderungen an arbeitsmedizinisch-toxikologische Analysen – Stand der Technik. Dt. Ärztebl. 37, C1753–C1760 (1997)
[30] J. Angerer, T. Göen and G. Lehnert: Mindestanforderungen an die Qualität von umweltmedizinisch-toxikologischen Analysen. Umweltmed. Forsch. Prax. 3, 307–312 (1998)
[31] R. A. Floyd, J. J. Watson, P. K. Wong, D. H. Altmiller and R. C. Rickard: Hydroxyl free radical adduct of deoxyguanosine: Sensitive detection and mechanisms of formation. Free Radic. Res. Commun. 1, 163–172 (1986)
[32] K. C. Cundy, R. Kohen and B. N. Ames: Determination of 8-hydroxydeoxyguanosine in human urine: A possible assay for in vivo oxidative DNA damage. Basic Life Sci. 49, 479–482 (1988)
[33] T. Renner, T. Fechner and G. Scherer: Fast quantification of the urinary marker of oxidative stress 8-hydroxy-2'-deoxyguanosine using solid-phase extraction and high-performance liquid chromatography with triple-stage quadrupole mass detection. J. Chromatogr. B Biomed. Sci. Appl. 738(2), 311–317 (2000)

Authors: *J. Lewalter, G. Leng, M. Wimber*
Examiners: *G. Scherer, T. Renner*

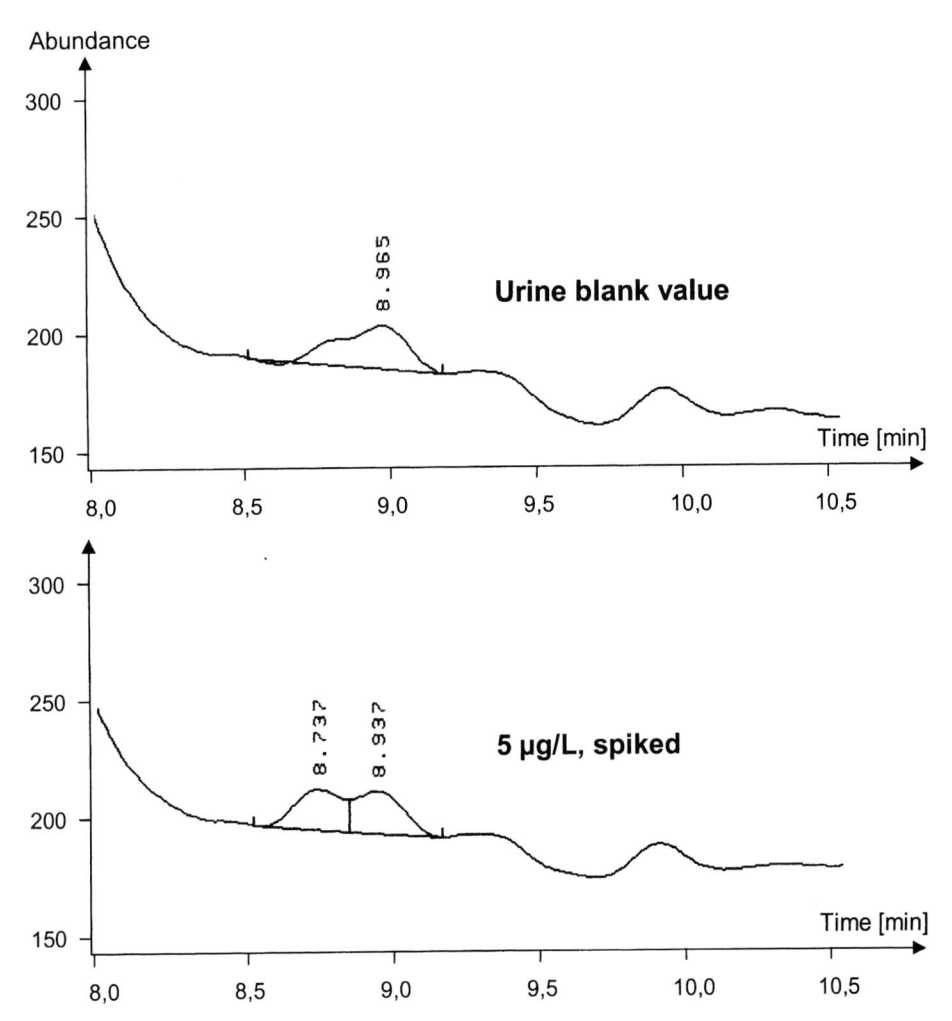

Fig. 1. Example chromatograms of a processed sample of pooled urine

8-Hydroxy-2′-deoxyguanosine

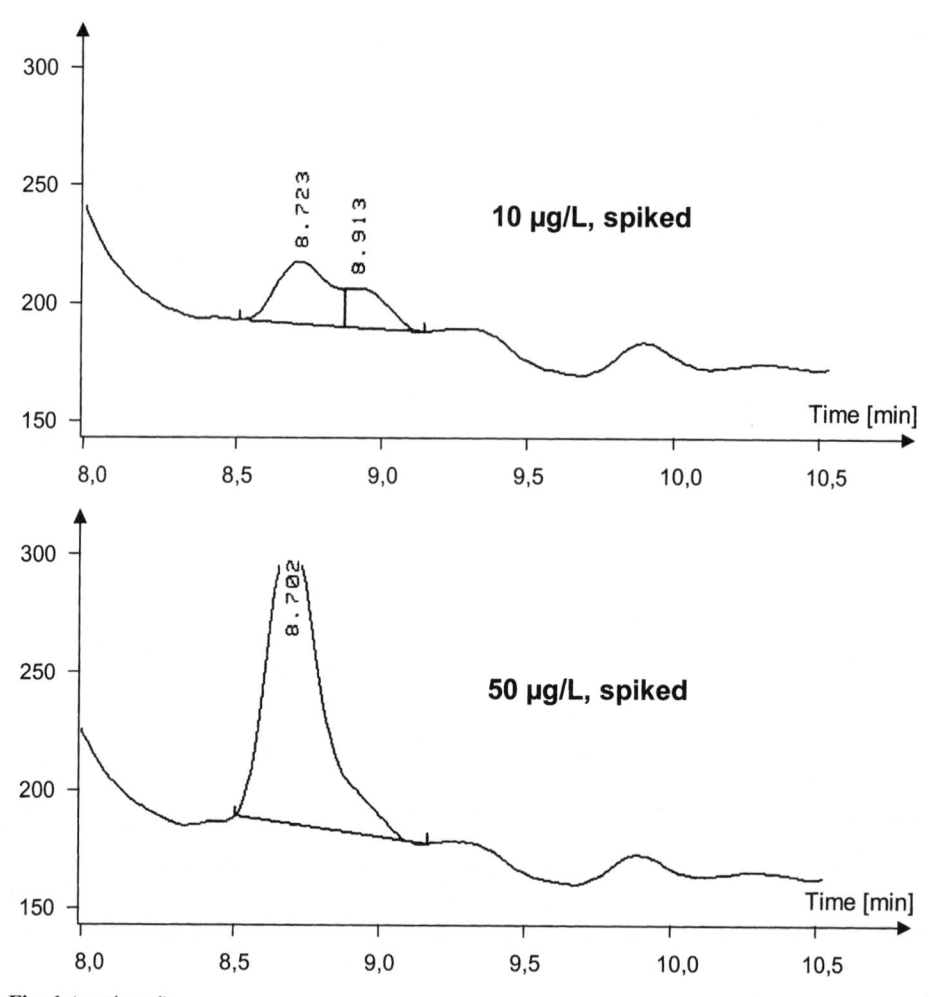

10 µg/L, spiked

8.723
8.913

50 µg/L, spiked

8.702

Fig. 1 (continued)

1-(4-(1-Hydroxy-1-methylethyl)-phenyl)-3-methylurea (HMEPMU) as a metabolite of isoproturon

Application   Determination in urine

Analytical principle Capillary gas chromatography/
                         mass spectrometric detection (MS)

Completed in   May 2001

Summary

Using this method, internal levels of isoproturon can be measured in persons who are exposed to the plant protection agent at work or in the environment. The main metabolite of isoproturon, 1-(4-(1-hydroxy-1-methylethyl)-phenyl)-3-methylurea (HMEPMU), is determined in urine by means of GC/MS-SIM. After alkaline hydrolysis to eliminate the methylcarbamide part of HMEPMU, the remaining 4-(1-hydroxy-1-methylethyl)-aniline (HMEA) is extracted using a ternary solvent mixture. The extract is dried with sodium sulphate. Derivatisation of HMEA is achieved by adding pentafluoropropionic acid anhydride to the organic phase and allowing it to react for 30 minutes at room temperature. The solution is then washed with phosphate buffer to remove any acid and evaporated in a stream of nitrogen. The analytical determination of the HMEA derivative is carried out by means of GC/MS in the selected ion monitoring mode. Reference standards prepared in pooled urine are used for calibration, and 4-(2-hydroxyethyl)-aniline is added to native urine as an internal standard.

Isoproturon

Within-series imprecision: Standard deviation (rel.) $s = 2.8\%$
                            Prognostic range (rel.) $u = 6.3\%$
                            at an HMEPMU concentration of 10 µg/L
                            and where $n = 9$ determinations

1-(4-(1-Hydroxy-1-methylethyl)-phenyl)-3-methylurea (HMEPMU)

Between-day imprecision: Standard deviation (rel.) $s = 4.7\%$
                                      Prognostic range (rel.) $u = 10.1\%$
                                      at an HMEPMU concentration of 10 µg/L
                                      and where $n = 14$ determinations

Quantification limit: 1.5 µg/L

Isoproturon

Isoproturon (N-(4-isopropylphenyl)-N,N'-dimethylurea, [CAS Number 34123-59-6]) is a urea-based plant protection agent. The crystalline compound is colourless and odourless. Its melting point is approx. 155 °C and its solubility in water is low (0.07 g/L). It has an n-octanol/water distribution coefficient (log P_{OW}) of 2.5.

Isoproturon has been officially registered as a herbicide in Germany for about 25 years. It is marketed alone or in combination with other active substances under various trade names, e.g. Arelon, Belgran, Blevigor, Dizan, Foxpro, Foxtar, Graminon, Herbatox, Hora-Turon, Longin and Tolkan [1].

After oral intake, isoproturon is rapidly metabolised. The main reactions are demethylation of the nitrogen atom and hydroxylation of the isopropyl group in the aromatic ring (Figure 1). The metabolite 1-(4-(1-hydroxy-1-methylethyl)-phenyl)-3-methylurea (HMEPMU) with a half-life of approximately 8 hours is excreted in the urine and represents about 50% of the absorbed isoproturon dose [2].

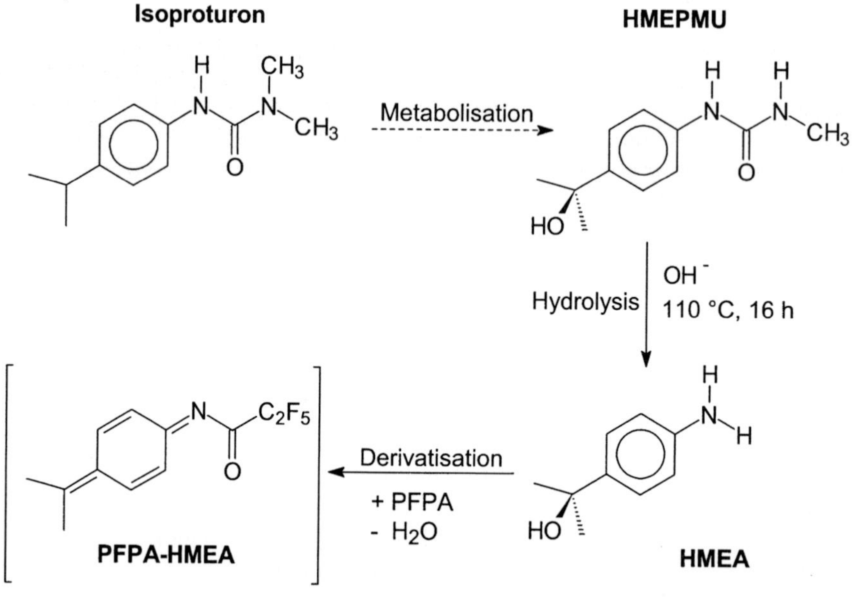

Fig. 1. Structures of isoproturon, its metabolite HMEPMU and the derivative formed under the conditions for sample processing

The only results published on the toxicology of isoproturon were obtained from animal studies, which mainly stem from the procedure to register the substance as a plant protection agent [1, 3–5]. According to these results, the acute, sub-chronic and chronic toxicity of isoproturon is low (e.g. rat: LD_{50} oral 1800–2400 mg/kg; mouse: LD_{50} oral 3350 mg/kg; rabbit: no skin or eye irritation; guinea pig: no skin sensitisation). Indications of a toxic effect on the liver can be observed above a dose of 400 mg/kg of food (male rats) or 2000 mg/kg of food (female rats). The weight of the liver increases and the incidence of liver tumours rises. Isoproturon is therefore regarded as a tumour-promoting substance [6]. At present there is no indication of any neurotoxic or foetotoxic attributes, teratogenity or mutagenity.

Based on the NOEL (No Observed Effect Level) for rats of 3.1 mg/kg body weight/day and taking a safety factor of 1000 into account the former Deutsche Bundesgesundheitsamt (German Ministry of Health) specified that a total daily intake (TDI) of 2.5 µg/kg body weight was tolerable for humans [7].

To date human exposure to isoproturon has only been described after an accidental release of the substance as a result of a malfunction in production [2, 8]. In this case the arithmetical mean levels in the urine of 168 cleaning personnel in three different highly contaminated areas of the affected production plant were 559 µg/L, 53 µg/L and 52 µg/L. No isoproturon was detected in 299 of the 454 samples investigated [2]. A timely investigation of the population living in the vicinity of the production plant (149 children, 5 adults) and 90 fire-fighters involved in the cleaning-up procedures showed that the quantitation limit (1.5 µg/L) was exceeded in only 19 cases (12%) (children/local population) or in 13 cases (14%) (fire-fighters). A general background exposure to isoproturon is not detectable.

Authors: *M. Bader, T. Göen, J. Angerer*
Examiner: *G. Müller*

1-(4-(1-Hydroxy-1-methylethyl)-phenyl)-3-methylurea (HMEPMU) as a metabolite of isoproturon

Application Determination in urine

Analytical principle Capillary gas chromatography/ mass spectrometric detection (MS)

Completed in May 2001

Contents

1   General principles
2   Equipment, chemicals and solutions
2.1 Equipment
2.2 Chemicals
2.3 Solutions
2.4 Calibration standards
3   Specimen collection and sample preparation
3.1 Sample preparation
4   Operational parameters
4.1 Operational parameters for gas chromatography and mass spectrometry
5   Analytical determination
6   Calibration
7   Calculation of the analytical result
8   Standardisation and quality control
9   Reliability of the method
9.1 Precision
9.2 Accuracy
9.3 Detection limits
9.4 Sources of error
10   Discussion of the method
11   References

1-(4-(1-Hydroxy-1-methylethyl)-phenyl)-3-methylurea (HMEPMU)

1 General principles

Using this method, internal levels of isoproturon can be measured in persons who are exposed to the plant protection agent at work or in the environment. The main metabolite of isoproturon, 1-(4-(1-hydroxy-1-methylethyl)-phenyl)-3-methylurea (HMEPMU), is determined in urine by means of GC/MS-SIM. After alkaline hydrolysis to eliminate the methylcarbamide part of HMEPMU, the remaining 4-(1-hydroxy-1-methylethyl)-aniline (HMEA) is extracted using a ternary solvent mixture. The extract is dried with sodium sulphate. Derivatisation of HMEA is achieved by adding pentafluoropropionic acid anhydride to the organic phase and leaving it to react for 30 minutes at room temperature. The solution is then washed with phosphate buffer to remove any acid and evaporated in a stream of nitrogen. The analytical determination of the HMEA derivative is carried out by means of GC/MS in the selected ion monitoring mode. Reference standards prepared in pooled urine are used for calibration, and 4-(2-hydroxyethyl)-aniline is added to native urine as an internal standard.

2 Equipment, chemicals and solutions

2.1 Equipment

Gas chromatograph/mass spectrometer with split injection device and the possibility of selected ion detection, automated injection system and an integration system or data processing system

Capillary column DB-17 (50%-phenyl)-methylpolysiloxane), length 60 m; inner diameter 0.25 mm; film thickness 0.2 µm (e.g. from J&W)

Glass beakers (250 mL, 1000 mL) (e.g. from Brand)

Glass and plastic vessels for storage of chemicals (e.g. from Brand)

Laboratory shaker (e.g. from Ikamag)

Laboratory centrifuge (e.g. Megafuge, from Heraeus)

Solvent evaporator (e.g. ReactiVap from Pierce)

Magnetic stirrer (e.g. from Ika)

Volumetric flasks (100 mL, 1000 mL) (e.g. from Brand)

pH meter with a single-rod electrode (e.g. from Mettler-Toledo)

Sample vials (1.8 mL) (e.g. from Macherey & Nagel)

Crimp-top vials (5 mL) (e.g. from Macherey & Nagel)

Screw-top tubes (10 mL, 20 mL) (e.g. from Brand)

Drying oven or equivalent heating device (e.g. from Heraeus)

Pipettes with variable volumes (10–100 µL, 100–1000 µL) (e.g. from Eppendorf)

1-(4-(1-Hydroxy-1-methylethyl)-phenyl)-3-methylurea (HMEPMU)

2.2 Chemicals

If not otherwise specified, all the chemicals used must be at least p.a. grade.

1-(4-(1-Hydroxy-1-methylethyl)-phenyl)-3-methylurea (HMEPMU)
(e.g. from AgrEvo, Frankfurt)

4-(2-Hydroxyethyl)-aniline (e.g. from Fluka)

Bidistilled water (equivalent to ASTM type I)

Di-potassium hydrogen phosphate (e.g. from Sigma-Aldrich)

Dichloromethane (e.g. from Merck)

Ethanol (e.g. from Merck)

n-Heptane (e.g. from Merck)

Sodium hydroxide (e.g. from Merck)

Sodium sulphate, anhydrous (e.g. from Merck)

ortho-Phosphoric acid (43%) (e.g. from Merck)

Pentafluoropropionic acid anhydride (e.g. from Fluka)

Hydrochloric acid (37%) (e.g. from Merck)

Toluene (e.g. from Merck)

2.3 Solutions

30% NaOH:
Approximately 50 mL bidistilled water are placed in a 250 mL glass beaker. After the addition of 30 g sodium hydroxide, the solution is mixed with a magnetic stirrer until it is clear, and then the volume is made up to a total of 100 mL with bidistilled water.

0.1 M Hydrochloric acid:
1 mL of 37% HCl is placed into a 100 mL volumetric flask and the flask is filled to its nominal volume with bidistilled water.

1 M Phosphate solution:
174 g of K_2HPO_4 are weighed into a 1000 mL volumetric flask and the flask is filled to its nominal volume with bidistilled water.

0.01 M Phosphate buffer (pH 8):
10 mL of 1 M phosphate solution are placed in a glass beaker and the beaker is filled to a total volume of 1000 mL with bidistilled water. The pH value is adjusted to 8 by adding *ortho*-phosphoric acid, and this is checked using a pH meter.

Ternary solvent mixture:
90 mL dichloromethane, 90 mL heptane and 20 mL toluene are mixed in a 250 mL glass beaker.

Solutions of the internal standard

Starting solution:
Approximately 50 mg 4-(2-hydroxyethyl)-analine are weighed exactly into a 100 mL volumetric flask. The flask is subsequently filled to its nominal volume with 0.1 M hydrochloric acid (500 mg/L).

Stock solution:
0.5 mL of the starting solution of the internal standard are pipetted into a 100 mL volumetric flask. The flask is subsequently filled to its nominal volume with bidistilled water (2.5 mg/L).

2.4 Calibration standards

Calibration standards and quality control solutions are prepared in pooled urine from test subjects who exhibit no known exposure to isoproturon. For the purpose of preparing pooled urine, spontaneous urine samples from the test subjects are collected in a suitable vessel, thoroughly mixed and stored at −18 °C until the standards and the control material are prepared.
All starting and stock solutions as well as calibration standards are stored in sealed vessels at approx. −18 °C, and they are stable under these conditions for at least 6 months.

Starting solution:
Approx. 50 mg 1-(4-(1-hydroxy-1-methylethyl)-phenyl)-3-methylurea (HMEPMU) are exactly weighed into a 100 mL volumetric flask. The flask is subsequently filled to its nominal volume with 0.1 M hydrochloric acid (500 mg/L).

Working solution:
1 mL of the starting solution is pipetted into a 100 mL volumetric flask. The flask is then filled to its nominal volume with pooled urine (5 mg/L).

Stock solution:
10 mL of the working solution are pipetted into a 100 mL volumetric flask. The flask is subsequently filled to its nominal volume with pooled urine (500 µg/L).
The calibration standards in pooled urine are prepared in accordance with the following pipetting scheme (Table 1).

Table 1. Pipetting scheme for the preparation of the calibration standards

Volume of the stock solution [mL]	Final volume of the calibration standard in urine [mL]	Concentration of the calibration standard [µg/L]
0	100	0
1	100	5
2	100	10
5	100	25
10	100	50

1-(4-(1-Hydroxy-1-methylethyl)-phenyl)-3-methylurea (HMEPMU)

In each case the calibration standards are divided into aliquots of 6 mL in sealable sample vials and stored in the deep-freezer at approx. −18 °C until processing. The calibration standards can be stored in this manner for at least 8 weeks.

3 Specimen collection and sample preparation

The urine samples are collected in sealable polyethylene bottles immediately after the end of a working shift and stored in the deep-freezer at approx. −18 °C until processing. The urine can be stored in this manner for at least six months.

3.1 Sample preparation

Before analysis, the samples are thawed in a water bath at 40 °C and thoroughly mixed. 10 mL of urine are pipetted into a 20 mL screw-top tube, 6 mL of 30% NaOH and 100 µL of the internal standard are added, and the tightly sealed tube is incubated for 16 hours at 110 °C in the drying cupboard. After cooling, the aqueous phase is extracted twice using 3 mL of the ternary solvent mixture in each case (laboratory shaker, 10 min, setting = 200 vibrations/min). The phase separation is then carried out by centrifugation (5 min, 1500 g). The upper organic phase is withdrawn after each extraction step using a 5 mL pipette, and it is transferred to a 10 mL screw-top tube. Approximately 50 mg anhydrous sodium sulphate are added to the combined extracts to dry them, and the tube is shaken vigorously (laboratory shaker, 5 min, setting = 200 vibrations/min). After centrifugation (5 min, 1500 g), the solution is transferred to a 10 mL screw-top tube, 50 µL pentafluoropropionic acid anhydride are added and the mixture is allowed to react for 30 min at room temperature. The solution is subsequently washed twice with 2 mL phosphate buffer (pH 8) in each case (laboratory shaker, 1 min, setting = 200 vibrations/min), centrifuged (5 min, 1500 g), withdrawn using a pipette and transferred to a 5 mL crimp-top vial. The organic phase is first evaporated to approximately 1 mL in a stream of nitrogen and, after transferring the contents anew to a 1.8 mL sample vial, it is further evaporated to a volume of 100 µL.

4 Operational parameters

4.1 Operational parameters for gas chromatography and mass spectrometry

Capillary column: Material: Fused silica
                   Stationary phase: DB-17
                   Length: 60 m
                   Inner diameter: 0.25 mm
                   Film thickness: 0.2 µm

1-(4-(1-Hydroxy-1-methylethyl)-phenyl)-3-methylurea (HMEPMU)

Detector:	Mass selective detector (MSD)	
Temperatures:	Column:	Initial temperature 80 °C, then increase at a rate of 5 °C/min to 180 °C, then at a rate of 20 °C/min to 280 °C, then 20 min at the final temperature
	Injector:	260 °C
	Transfer line:	280 °C
Carrier gas:	Helium 5.0 with a constant flow of 1 mL per minute	
Split:	Splitless	
Sample volume:	1 µL	
Ionisation type:	Electron impact ionisation (EI)	
Ionisation energy:	70 eV	
Dwell time:	100 ms	
Electron multiplier:	2800 V (abs.)	

All other parameters must be optimised in accordance with the manufacturer's instructions.

5 Analytical determination

In each case 1 µL is injected into the gas chromatograph for the analytical determination of the urine samples processed as described in Section 3.1. A quality control sample is analysed with each analytical series.

The temporal profiles of the ion traces shown in Table 2 are recorded in the SIM mode.

Table 2. Retention times and masses

Compound	Retention time [min]	Masses
Isoproturon (HMEA derivative)	17.9	279* 264 132
4-(2-Hydroxyethyl)-aniline (IS)	15.6	429*

The masses marked with * are used for quantitative evaluation.

The retention times shown in Table 2 serve only as a guide. Users of the method must satisfy themselves of the separation power of the capillary column used and the resulting retention behaviour of the substances. Figure 2 shows an example of a chromatogram of a calibration standard with an isoproturon concentration of 5 µg/L.

6 Calibration

The calibration standards are processed in the same manner as the urine samples (Section 3.1) and analysed by means of gas chromatography/mass spectrometry as described in Sections 4 and 5. Linear calibration graphs are obtained by plotting the quotients of the peak areas of the isoproturon metabolite derivative and that of the internal standard as a function of the concentrations used. It is unnecessary to plot a complete calibration graph for every analytical series. It is sufficient to analyse one calibration standard for every analytical series. The ratio of the result obtained for this standard and the result for the equivalent standard in the complete calibration graph is calculated. Using this quotient, each result read off the calibration graph is adjusted for the relevant series. A typical linear calibration graph is shown in Figure 3.
New calibration graphs should be plotted if the quality control results indicate systematic deviation.
The calibration graph is linear between the detection limits and 200 µg per litre urine.

7 Calculation of the analytical result

Quotients are calculated by dividing the peak areas of the analyte by the peak area of the internal standard. These quotients are used to read off the pertinent concentration of the HMEPMU derivative in µg per litre from the relevant calibration graph.

8 Standardisation and quality control

The guidelines given by the Bundesärztekammer (German Medical Association) [9, 10] and the special preliminary remarks to this series are to be followed to ensure the quality of the analytical results. In order to determine the precision of the method, a urine sample containing a constant concentration of HMEPMU is analysed. As material for quality control is not commercially available, it must be prepared in the laboratory. For this purpose, urine is spiked with a defined quantity of HMEPMU. The concentration of this control material should lie within the appropriate concentration range. A six-month supply of the control material is prepared, divided into aliquots and stored in the deep-freezer. The theoretical value and the tolerance range for this quality control material are determined in a preliminary period (one analysis of the control material on each of 20 different days) [11, 12].

9 Reliability of the method

9.1 Precision

The precision in the series was determined by processing and analysing control material with a spiked HMEPMU concentration of 10 µg/L. When these urine samples were determined in 9 replicate determinations, the relative standard deviation of the results was 2.8%, equivalent to a prognostic range of 6.3% (cf. Table 3).

Table 3. Precision for the determination of the isoproturon metabolite

	n	Concentration [µg/L]	Standard deviation (rel.) [%]	Prognostic range [%]
In the series	9	10	2.8	6.3
From day to day	14	10	4.7	10.1

In addition, the precision from day to day was determined. The same material as for the determination of the precision in the series was used for this purpose. This urine was processed and analysed on each of 14 different days. The standard deviation was 4.7%, which is equivalent to a prognostic range of 10.1% (cf. Table 3).

9.2 Accuracy

Recovery experiments were performed to check the accuracy of the method. For this purpose pooled urine without HMEPMU and urine spiked with defined quantities of HMEPMU was analysed 9 times. The relative recovery rate was 93% at a concentration of 10 µg/L. The accuracy was not determined by measuring the losses due to processing, as no appropriate reference substances were available.

9.3 Detection limits

Reliable determination of the actual detection limit of the method in urine was impossible on account of an interfering peak which occurred in almost all the samples. The area of the interfering peak in pooled urine (N = 12 test subjects) was equivalent to a concentration of about 0.5 µg/L on average. A concentration of 1.5 µg/L was therefore determined as the quantitation limit, which was taken as three times the ratio of the analyte signal to the interfering signal.

9.4 Sources of error

It is extremely important for reproducible hydrolysis of HMEPMU that the drying oven is maintained at the correct temperature and that the reaction is allowed to proceed for the required length of time. In the experience of the authors, not every drying oven displays the actual temperature correctly; therefore it is advisable to check the temperature using an external thermometer.

The quantification of the HMEPMU derivative is often impeded by an interfering peak at almost the same retention time (Figure 2, Figure 4). The peak is not a reagent blank value, but an unidentified component in the urine. As in the case of the HMEA derivative, the interfering peak exhibits fragments with m/z = 132 and 279. To date neither gas chromatographic separation nor mass spectrometric fade-out has been successful in eliminating the interference, even after variation of the normal instrumental parameters (temperatures, column, pressure).

10 Discussion of the method

Biomonitoring after occupational exposure to isoproturon is normally carried out by means of HPLC-UV [2] in the practice of occupational medicine. However, this method is not suitable for investigations at low dosage levels on account of the relatively high detection limit of 50 µg/L. The GC/MS method described here achieves a considerably lower quantitation limit of 1.5 µg/L, which is primarily limited by an unidentified interfering component. This interfering component is present in the urine of the general population at concentrations of less than 0.5 µg/L. It is assumed for the sake of simplicity that this interfering component is detected with the same sensitivity as the analyte. The method suggested here can therefore be used for biological monitoring despite the interference, as any peak that is equivalent to a concentration of more than 1.5 µg/L is caused by isoproturon with an error probability of 5%.

It was established during the development of the method that it is not the expected pentafluoropropionic acid derivative of 4-(1-hydroxy-1-methylethyl)-aniline (HMEA) that is analysed in the last instance, but the derivative of a dehydration product presumably with a quinoid structure, 1-imino-4-isopropylenyl-cyclohexa-2,5-diene (see also Figure 1 and Figure 5). It was not established whether this compound is formed by elimination of water during the alkaline hydrolysis or only by heating the derivative in the GC/MS injection system. The good reproducibility of the method and its confirmed robustness, e.g. with respect to matrix effects, makes the use of this method justified, although the mechanisms that lead to the formation of the analyte have not been clearly characterised to date.

The analytical reliability criteria of the method were checked using control material prepared in the laboratory. Pooled urine from persons who were not exposed to isoproturon was spiked with a concentration of 10 µg/L HMEPMU and divided into aliquots. The precision in the series of approx. 3% and the precision from day to day of about 5% confirm the reproducibility and the robustness of the method. It proved impossible to determine the accuracy of the procedure, as the necessary reference substances were not available.

Calibration of the method using aqueous standards is not possible. The recovery of the HMEA derivative is extremely low in this case, as measurement signals are only obtained above a concentration of 100 µg/L.

To summarise, this method is very suitable for investigating the inner exposure of persons who have handled this substance in the course of their work. Exposure of the general population, as in the case of the pyrethroids or organophosphates for example, has not been detected to date. Therefore this method is useful only if it has been established that the test subject has had contact with isoproturon.

Instruments used:
Gas chromatograph Hewlett-Packard GC 5890 Series II plus, mass spectrometer Hewlett-Packard MSD 5972, autosampler Hewlett-Packard 7673 A.

11 References

[1] *Industrieverband Agrar (ed.):* Wirkstoffe in Pflanzenschutz- und Schädlingsbekämpfungsmitteln, 2nd, newly revised edition. BLV Verlagsgesellschaft mbH, Munich, Vienna, Zürich (1990)
[2] *R. Kaleja:* Arbeitsmedizinische und toxikologische Erfahrungen aus einer umweltrelevanten Betriebsstörung. Arbeitsmedizin, Sozialmedizin, Umweltmedizin 32, 208–211 (1997)
[3] *K.H. Leist and E. Ebert:* Isoproturon – Summary and evaluation of the toxicological data for isoproturon – active ingredient technical (Hoe 016410=CGA 18731=35689 R.P.). Hoechst AG, unpublished investigation, Report No. 90.0962 (1990)
[4] *K.H. Leist:* Isoproturon: Evaluation of the initiation and/or promotion activity. Hoechst AG, unpublished investigation, DOC A 44049 (1990)
[5] *K.H. Leist:* Isoproturon: Reevaluation of the initiation and/or promotion activity. Hoechst AG, unpublished investigation. Amendment to DOC A 44049 (1992)
[6] *World Health Organization:* Guidelines for drinking-water quality, 2nd edition, Volume 1 Recommendations: Isoproturon. WHO, Geneva (1993)
[7] *BGA (Bundesgesundheitsamt):* ADI-Werte und DTA-Werte für Pflanzenschutzmittelwirkstoffe. 3rd edition. Bundesgesundheitsblatt 27, 250 (1993)
[8] *U. Heudorf, M. Bader, A. Koch, U. Ewers and J. Angerer:* Humanbiomonitoring: Expositions- und Risikoabschätzung nach einem Chemieunfall. Umweltmed. Forsch. Prax. 1, 23–26 (1997)
[9] *Bundesärztekammer:* Qualitätssicherung der quantitativen Bestimmungen im Laboratorium. Neue Richtlinien der Bundesärztekammer. Dt. Ärztebl. 85, A699–A712 (1988)
[10] *Bundesärztekammer:* Ergänzung der „Richtlinien der Bundesärztekammer zur Qualitätssicherung in medizinischen Laboratorien". Dt. Ärztebl. 91, C159–C161 (1994)
[11] *G. Lehnert, J. Angerer and K.H. Schaller:* Statusbericht über die externe Qualitätssicherung arbeits- und umweltmedizinisch-toxikologischer Analysen in biologischen Materialien. Arbeitsmed. Sozialmed. Umweltmed. 33(1), 21–26 (1998)
[12] *J. Angerer and G. Lehnert:* Anforderungen an arbeitsmedizinisch-toxikologische Analysen – Stand der Technik. Dt. Ärztebl. 37, C1753–C1760 (1997)

Authors: *M. Bader, T. Göen, J. Angerer*
Examiner: *G. Müller*

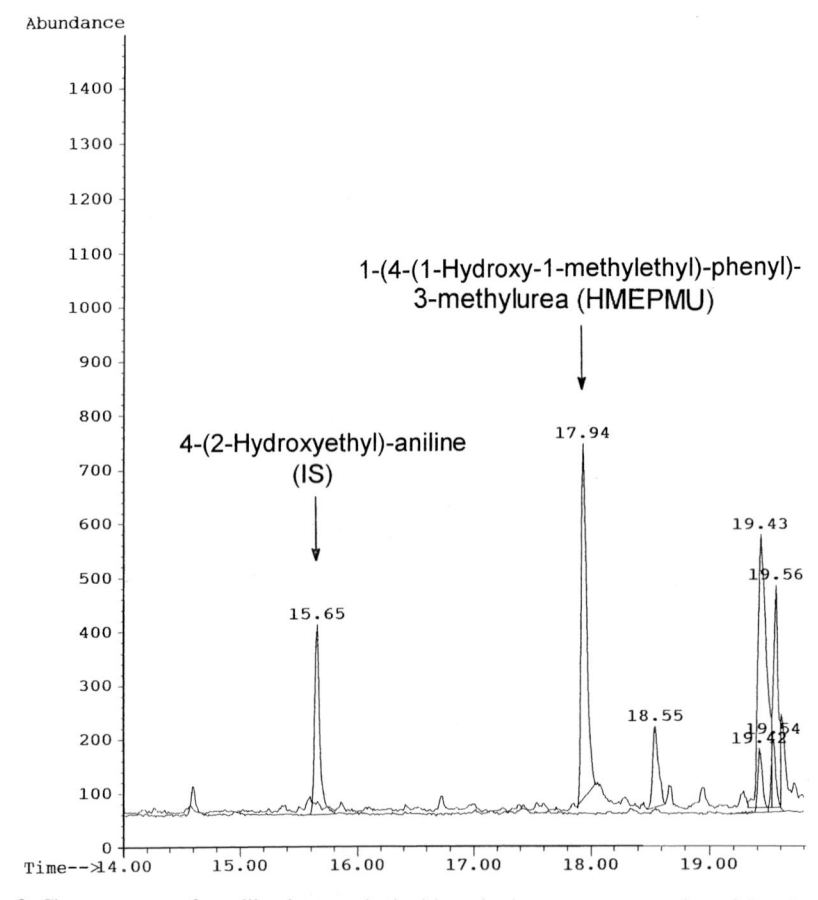

1-(4-(1-Hydroxy-1-methylethyl)-phenyl)-3-methylurea (HMEPMU)

Fig. 2. Chromatogram of a calibration standard with an isoproturon concentration of 5 µg/L

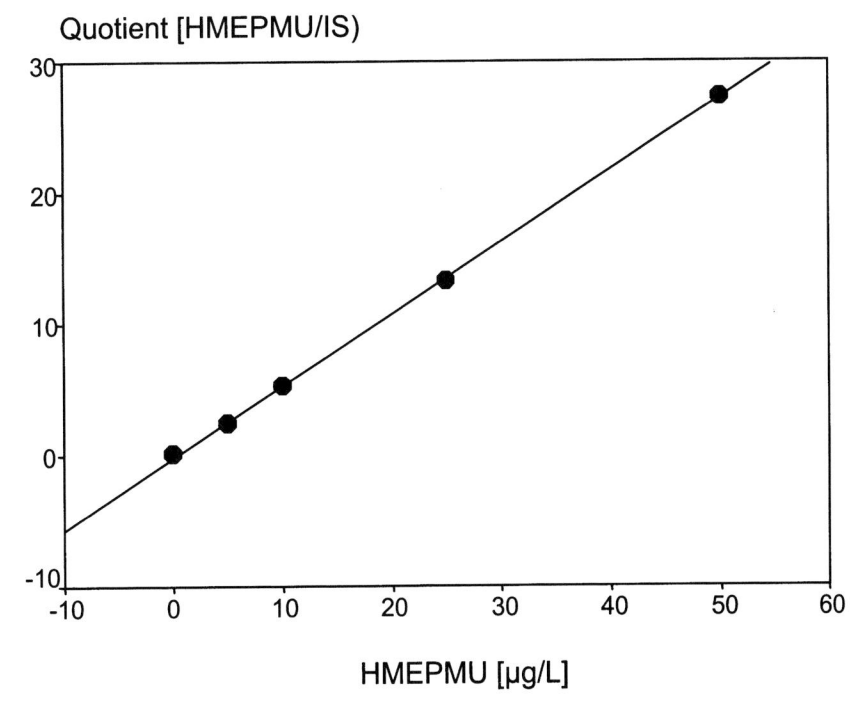

1-(4-(1-Hydroxy-1-methylethyl)-phenyl)-3-methylurea (HMEPMU)

Fig. 3. Typical calibration function (pooled urine, y=0.547x − 0.176)

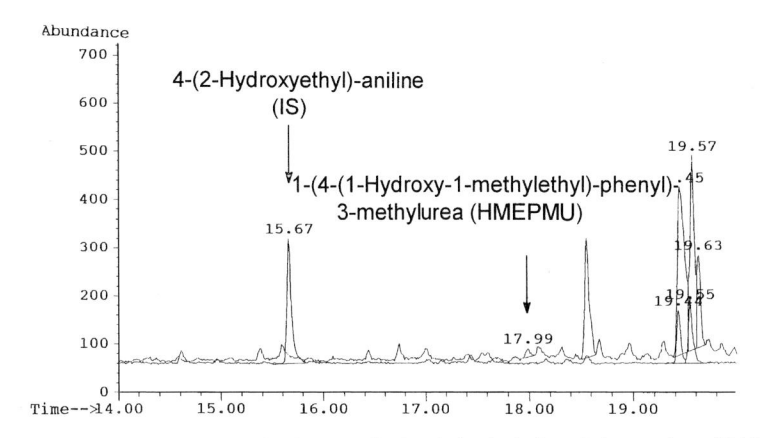

Fig. 4. Chromatogram of a processed urine sample (pooled urine), interfering peak at 17.99 minutes

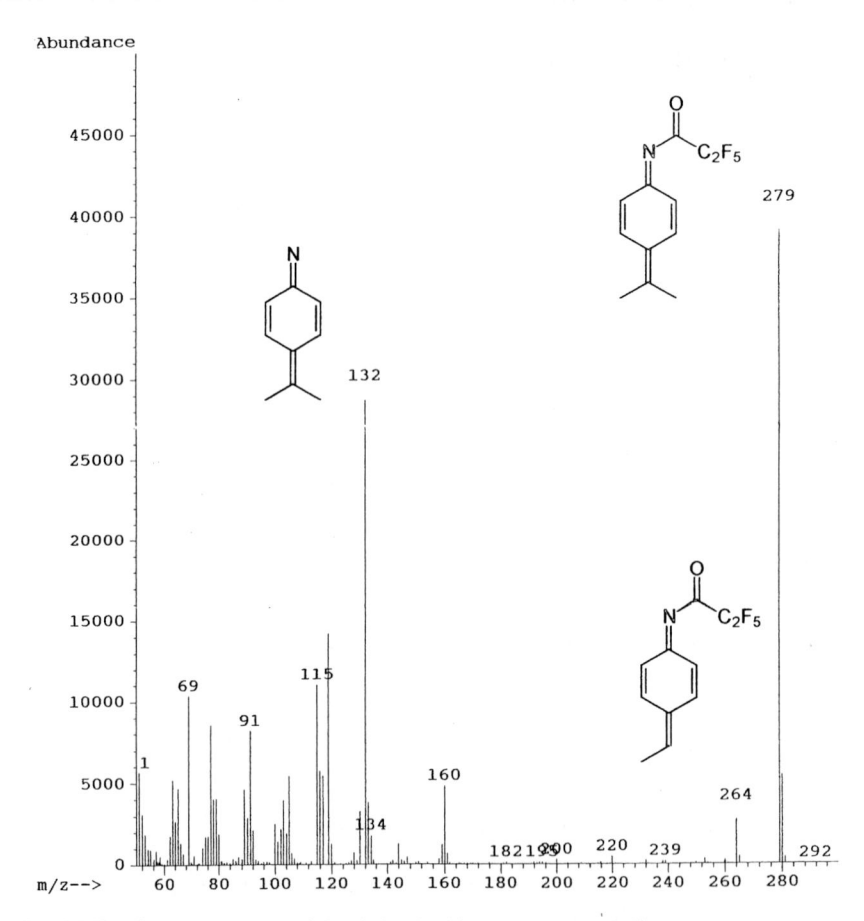

1-(4-(1-Hydroxy-1-methylethyl)-phenyl)-3-methylurea (HMEPMU)

Fig. 5. GC/MS reference spectrum of the derivatised isoproturon metabolite

Molybdenum

Application Determination in plasma/serum and urine

Analytical principle Inductively coupled plasma-quadrupole mass spectrometry with electrothermal vaporisation

Completed in December 2001

Summary

Molybdenum can be simply, sensitively and specifically determined in plasma/serum or urine using the ICP-MS method described here. The procedure is suitable for detecting molybdenum at levels down to the physiological concentration range. The plasma or urine samples are vaporised electrothermally in a graphite furnace without additional digestion of the samples and then introduced into the ICP-MS. Quantification is carried out using the standard addition procedure.

Molybdenum in plasma/serum

Within-series imprecision: Standard deviation (rel.) $s = 4.5\%$
Prognostic range $u = 10.0\%$
at a concentration of 0.6 µg Mo per litre plasma
and where $n = 10$ determinations

Between-day imprecision: Standard deviation (rel.) $s = 5.0\%$
Prognostic range $u = 11.1\%$
at a concentration of 0.6 µg Mo per litre plasma
and where $n = 10$ determinations

Accuracy: Recovery rate: $r = 97–103\%$ at a concentration of 1 µg Mo per litre plasma

Detection limit: 0.05 µg molybdenum per litre plasma

Molybdenum in urine

Within-series imprecision: Standard deviation (rel.) $s = 2.0\%$
Prognostic range $u = 4.5\%$
at a concentration of 12 µg Mo per litre urine
and where $n = 10$ determinations

Between-day imprecision: Standard deviation (rel.) $s = 3.0\%$
Prognostic range $u = 6.7\%$
at a concentration of 12 µg Mo per litre urine
and where $n = 10$ determinations

Accuracy: Recovery rate: $r = 98–104\%$
at a concentration of 10 µg Mo per litre urine

Detection limit: 0.05 µg molybdenum per litre urine

Molybdenum

Molybdenum (Mo) has an atomic number of 42 and an atomic weight of 95.94. Metallic Mo has a density of 10.2 g/cm^3, melts at 2610 °C and boils at 5560 °C. Molybdenum occurs most commonly in tetravalent (e.g. molybdenum sulphide) or hexavalent compounds, but also in bivalent, trivalent and pentavalent compounds.
According to recent estimates the Earth's crust contains about 1 ppm of molybdenum. The content can vary between 20 and 3000 ppm in different rock formations. The element is found in elevated concentrations in slate, phosphorite, coal and crude oil, whereby the content of molybdenum is directly proportional to that of organic carbon [1]. Molybdenum enters the environment as a result of smelting, through the use of phosphate-containing fertilisers and phosphate-containing detergents, and due to the combustion of coal and oil.
Annual world production of molybdenum is about 80,000 t (mainly from molybdenum sulphide, the largest amounts are produced in the USA). It is used for steel production (structural steel, stainless steel, tool steel, see also [2]) and in special chemical products (paint pigment, catalysts, lubricants and flame-retardants) [3].
Molybdenum is an essential trace element. It is readily absorbed from the gastrointestinal tract. The molybdenum content of food varies depending on the type of food. The highest concentrations of molybdenum are found in milk products, dried leguminous vegetables, innards (liver, kidneys), grain products and baked goods. The lowest levels are found in pork, fish, nuts, fat, sweets and citrus fruits [4, 5]. The molybdenum content of plants is strongly dependent on the relevant soil. Neutral or alkaline soil, especially with a high moisture content and a high concentration of organic materials, promotes the intake of the element [6, 7]. Drinking water with a content of 0 to 20 µg/L contributes very little to dietary uptake of molybdenum. Values of the order of ten times higher were determined in highly industrialised areas with high levels of molybdenum in the soil, e.g. in some parts of Colorado where the concentration is about 400 µg/L, supposedly due to the effluent from mines [5].
In 1980 the National Research Council of the USA considered a daily intake of 50 to 100 µg for children from 1 to 3 years and 150 to 500 µg for adults to be sufficient [8]. Apparently the intake in New Zealand is considerably below this range given for US citizens [9].
The absorbed molybdenum is mainly excreted in the urine, but partly also with bile [10, 11]. Studies were carried out on the metabolic balance in 36 girls between the

ages of 6 and 10 years [12] and in 4 young New Zealand women (19 to 21 years old) [13]. The intake was estimated at values between approximately 45 and 80 or 95 µg/day in these investigations. The figures [12] show that absorption of the element fluctuates considerably depending on the diet.

Iversen and co-workers [14] determined reference values for the Danish general population as part of the EURO-TERVIHT Project (Trace Element Reference Values in Human Tissues). Investigation of 128 persons using ICP-MS showed a median excretion of 42 µL molybdenum per litre urine.

In 1999 the urine of 904 persons from the general population in the USA without prior known occupational exposure to the element was tested for molybdenum as part of the NHANES Programme (National Health and Nutrition Examination Survey). The results showed a median renal excretion level of 53 µg Mo per litre urine. The 90th percentile was about 140 µg/L [15].

Molybdenum concentrations in the plasma of persons (n = 30) who were not exposed to this metal at the workplace are 0.6 µg per litre plasma on average [16]. These concentrations are in very good agreement with the plasma values already measured in the general population in the 80s [17–21].

The extremely high incidence of endemic gout in Ankavan Province in Armenia in population groups that mainly live on the food that is locally grown there was associated with the abnormally high molybdenum concentrations in the soil and plants [22]. The molybdenum intake in the affected regions was estimated at 10 to 15 mg/day compared with 1 to 2 mg in nearby areas with a lower incidence of gout. Humans and grazing animals were reported to have abnormally high uric acid levels in the plasma and high activities of xanthine oxidase in the tissues.

In the case of animals the effect of a high intake of molybdenum with food depends on various factors: the age of the animal, the amount and the chemical form of the molybdenum, copper metabolism, and the phosphate, sulphur, protein, cysteine, methionine, zinc and lead content of the diet. Molybdenosis occurs in cattle in various parts of the world under normal grazing conditions. In Britain the disorder is known as "teart", in New Zealand it is known as "peatscours". The diarrhoea can occur in a mild form or as a severe illness that leads to permanent damage, and can be fatal. Sheep are less affected than cattle, and horses are not affected at all [6].

Walravens and co-workers [23, 24] investigated 25 workers who were employed in a molybdenum roasting furnace. The exposure to molybdenum dust was accompanied by slight increases in uric acid levels, a marked rise in the ceruloplasmin in the plasma, and a distinctly elevated molybdenum value in the plasma and urine. The copper content of the urine was highly elevated in 1 of 14 samples. There was no solid evidence of a molybdenum-related gout syndrome. Further studies are necessary to clarify any possible correlation with molybdenum.

The currently valid German MAK value for molybdenum and its water-insoluble compounds was stipulated as 4 mg/m^3, and as 5 mg/m^3 for molybdenum compounds that are soluble in water. To date no BAT value has been assigned [25].

Great advances have been made in research of the biochemical and functional significance of molybdenum. The element is evidently an essential component of the enzymes xanthine oxidase [26, 27], aldehyde oxidase [28] and sulphite oxidase [29]. To date cytoplasmic xanthine oxidase, which is active in the metabolism of purine, has been best investigated. The enzyme is important for the oxidation of hypoxanthine to xanthine and of xanthine to uric acid. Aldehyde oxidase has not yet been so thoroughly researched, but recently interest has been awakened due to its possible role in the metabolism of alcohol in the liver. Mitochondrial sulphite oxidase serves as the terminal enzyme in the decomposition of sulphur-containing amino acids. In bacteria and plants the reduction of nitrates and di-nitrogen compounds depends on the molybdenum content of nitrogenase and nitroreductase.

Johnson and co-workers [30] showed that aldehyde oxidase, sulphite oxidase, xanthine oxidase and nitrate reductase all contain the metal as a complex with an obviously identical organic component, which was isolated and identified as a central pteridine with an (as yet) unidentified 6-alkyl side chain [30–32]. Detailed reconstruction studies have shown that the molybdenum cofactor in these 3 enzymes is interchangeable from the functional point of view. Of the characteristic molybdenum enzymes, only nitrogenase contains a special iron-molybdenum cofactor.

Tungsten is obviously a competitive antagonist of molybdenum utilisation in the animal organism. Cohen and co-workers [33, 34] were able to show that the activity of sulphite oxidase and xanthine oxidase was reduced when tungsten was administered to rats. Considerable amounts of inactive protein of both enzymes were detected by specific antibodies against rat liver sulphite oxidase and xanthine oxidase in the liver of animals that had been treated with tungsten. Accordingly the biosynthesis proceeds rapidly, even when molybdenum utilisation is low. However, either the inactive enzymes contained tungsten instead of molybdenum or for the most part they contained no metal at all [35].

Information on the role played by molybdenum in human disorders is scarce. The plasma level was determined at the University of Ghent by means of neutron activation analysis of patients suffering from liver and biliary system disorders [21]. The accuracy and precision of the method was checked by the IAEA by means of analyses in Bowen's carbon powder, NBS bovine liver and dried animal blood [36, 37]. According to the results the molybdenum level in plasma was distinctly elevated during the initial phase of acute viral hepatitis (3.1 ± 1.46 ng/mL in patients; 0.55 ± 0.21 ng/mL in control persons) and sank together with the usual liver functional parameters to normal levels during recovery. A significant correlation was observed between the concentration of molybdenum in plasma and the plasma aspartate transaminase (SGOT) ($r = 0.710$; $p<0.001$) and the alanine transaminase (SGPT) ($r = 0.683$; $p<0.001$). In addition, distinctly elevated plasma molybdenum levels were found in a series of other hepatobiliary diseases, i.e. hepatitis B (Hb_sAg-positive) chronic active hepatitis, primary biliary cirrhosis, liver damage due to alcohol, medication-induced liver damage, liver metastases, gall stones, tumours of the gall bladder and the extra-hepatic bile ducts and carcinoma in the head of the pancreas. The mechanism responsible still remains unclear. There is obviously a close relationship between the uptake of circulating molybdenum in the liver and the functional state of the hepatocytes [38, 39]. A release of intracellular molybdenum from damaged parenchyma cells is also under consideration. As considerable increases of intra-extra-hepatic cholestase were also observed, interruption of the flow of bile may also be the cause. Further disorders for which elevated molybdenum values have been reported (uraemia, rheumatic disorders, neoplasia, cardiovascular diseases, infection) are given in the literature [40–45]. Only some of these results have been confirmed by analysis of reference material. Not all of the analysts quoted here have described sample collection and processing in detail. In such cases it remains uncertain if all the necessary precautions were taken to prevent exogenous contamination of the samples.

Author: *P. Schramel*
Examiners: *L. Dunemann, J. Begerow, P. Brätter*

Molybdenum

Application Determination in plasma/serum and urine

Analytical principle Inductively coupled plasma-quadrupole mass spectrometry with electrothermal vaporisation

Completed in December 2001

Contents

1 General principles
2 Equipment, chemicals and solutions
2.1 Equipment
2.2 Chemicals
2.3 Solutions
3 Specimen collection and sample preparation
3.1 Specimen collection and storage
4 Operational parameters
4.1 Plasma settings
4.2 Mass spectrometric parameters
4.3 Parameters for electrothermal vaporisation (ETV)
5 Analytical determination
6 Calibration and calculation of the analytical result
7 Standardisation and quality control
8 Evaluation of the method
8.1 Precision
8.2 Accuracy
8.3 Detection limits
8.4 Sources of error
9 Discussion of the method
10 References

1 General principles

The plasma or urine samples are vaporised electrothermally via an ETV system with a graphite furnace without additional digestion of the samples and then introduced into the ICP-MS. Quantification is carried out using the standard addition procedure.

2 Equipment, chemicals and solutions

2.1 Equipment

ICP mass spectrometer (quadrupole) with a device for electrothermal vaporisation and a graphite furnace, PC, autosampler and printer

Graphite tubes, pyrolytically coated

Laboratory centrifuge (e.g. from Heraeus)

Microlitre pipette, adjustable between 100 and 1000 µL (e.g. from Eppendorf)

Millilitre pipette, adjustable between 1 and 10 mL (e.g. from Eppendorf)

1000 mL Volumetric flask

10 mL Volumetric flask

10 mL Sample vessel made of plastic or quartz glass

10 mL Polyethylene tubes with stoppers (e.g. from Sarstedt)

EDTA Monovettes (e.g. from Sarstedt)

2 mL Plastic sample vessels for the AS 60 autosampler (from Perkin-Elmer)

2.2 Chemicals

Molybdenum standard solution, 1 g Mo/L in H_2O [$(NH_4)_6Mo_7O_{24} \cdot H_2O$] (e.g. from Spex)

Ultrapure water (e.g. Milli-Q, from Millipore)

35% Hydrochloric acid, ultrapure (e.g. Suprapur, from Merck)

$Pd(NO_3)_2$ Standard solution 20 g/L (e.g. from Spex)

$Mg(NO_3)_2$ Standard solution 20 g/L (e.g. from Spex)

65% HNO_3 (p.a.) subsequently purified by subboiling distillation

Argon auxiliary gas (welding argon or liquid argon) (e.g. from Messer-Griesheim)

Argon plasma gas (welding argon or liquid argon) (e.g. from Messer-Griesheim)

Argon nebulizer gas (welding argon or liquid argon) (e.g. from Messer-Griesheim)

2.3 Solutions

1 M HNO_3:
70 mL 65% HNO_3 are pipetted into a 1000 mL volumetric flask into which approx. 500 mL ultrapure water have already been placed, and the flask is filled to its nominal volume with ultrapure water while swirling the contents.
This solution can be stored for several months at 4 °C.

$Pd(NO_3)_2$ matrix modifier solution:
Approx. 6 mL 1 M nitric acid are placed into a 10 mL volumetric flask. 50 µL of the $Pd(NO_3)_2$ standard solution (20 g/L) are added, and the flask is filled to its nominal volume with 1 M nitric acid (100 mg/L).

$Mg(NO_3)_2$ matrix modifier solution:
Approx. 6 mL 1 M nitric acid are placed into a 10 mL volumetric flask. 100 µL of the $Mg(NO_3)_2$ standard solution (20 g/L) are added, and the flask is filled to its nominal volume with 1 M nitric acid (200 mg/L).

2.3.1 Standard solutions for the determination in plasma

Starting solution:
0.1 mL of the Mo standard solution (cf. Section 2.2) is pipetted into a 10 mL volumetric flask into which approx. 6 mL water has already been placed. Then 1 mL hydrochloric acid is added using a pipette. The flask is subsequently filled to its nominal volume with ultrapure water (10,000 µg/L).

Stock solution:
0.1 mL of the starting solution is pipetted into a 10 mL volumetric flask into which approx. 6 mL water has already been placed. Then 1 mL hydrochloric acid is added using a pipette. The flask is subsequently filled to its nominal volume with ultrapure water (100 µg/L).

Working solution:
1 mL of the stock solution is pipetted into a 10 mL volumetric flask into which approx. 6 mL water has already been placed. Then 1 mL hydrochloric acid is added using a pipette. The flask is subsequently filled to its nominal volume with ultrapure water (10 µg/L).
The stock solution and the working solution must be freshly prepared before each analytical run.
The starting solution can be stored for several (2 to 3) months in the refrigerator (at +4 °C).

Spiking solutions (plasma) for standard addition

Spiking solution 0:
Approx. 6 mL water are placed in a 10 mL volumetric flask. Then 1 mL hydrochloric acid is added using a pipette. The flask is subsequently filled to its nominal volume with ultrapure water (0 µg/L).

Spiking solution A:
1 mL of the working solution is pipetted into a 10 mL volumetric flask into which approx. 6 mL water has already been placed. Then 1 mL 35% hydrochloric acid is added using a pipette. The flask is subsequently filled to its nominal volume with ultrapure water (1 µg/L).

Spiking solution B:
2 mL of the working solution is pipetted into a 10 mL volumetric flask into which approx. 6 mL water has already been placed. Then 1 mL 35% hydrochloric acid is added using a pipette. The flask is subsequently filled to its nominal volume with ultrapure water (2 µg/L).
These spiking solutions are added to the plasma samples to be investigated (P0, P1, P2). For this purpose 0.1 mL of the appropriate spiking solution is added to 0.1 mL of each of the plasma samples in the autosampler vessels as shown in the pipetting scheme (Table 1), and the contents are thoroughly mixed. These sample solutions are then subjected to further sample processing.

Table 1. Pipetting scheme to prepare the spiked plasma sample solutions for the standard addition procedure and the resulting concentrations in µg/L

Designation	Volume of the spiking solutions			Volume of Plasma [mL]	Final volume [mL]	Spiked concentration [µg/L]
	0 [mL]	A [mL]	B [mL]			
P0	0.1	–	–	0.1	0.2	+ 0
P1	–	0.1	–	0.1	0.2	+ 0.5
P2	–	–	0.1	0.1	0.2	+ 1

2.3.2 Standard solutions for the determination in urine

Starting solution:
0.1 mL of the Mo standard solution (cf. Section 2.2) is pipetted into a 10 mL volumetric flask into which approx. 6 mL water had already been placed. Then 1 mL concentrated nitric acid is added using a pipette. The flask is subsequently filled to its nominal volume with ultrapure water (10,000 µg/L).

Working solution:
0.1 mL of the starting solution is pipetted into a 10 mL volumetric flask into which approx. 6 mL water had already been placed. Then 1 mL concentrated nitric acid is

added using a pipette. The flask is subsequently filled to its nominal volume with ultrapure water (100 µg/L).
The working solution must be freshly prepared before each analytical run.

Spiking solutions (urine) for standard addition

Spiking solution 0:
1 mL concentrated nitric acid is pipetted into a 10 mL volumetric flask into which approx. 6 mL water has already been placed. The flask is subsequently filled to its nominal volume with ultrapure water (0 µg/L).

Spiking solution A:
0.4 mL of the working solution is pipetted into a 10 mL volumetric flask into which approx. 6 mL water has already been placed. Then 1 mL concentrated nitric acid is added using a pipette. The flask is subsequently filled to its nominal volume with ultrapure water (4 µg/L).

Spiking solution B:
0.8 mL of the working solution is pipetted into a 10 mL volumetric flask into which approx. 6 mL water has already been placed. Then 1 mL concentrated nitric acid is added using a pipette. The flask is subsequently filled to its nominal volume with ultrapure water (8 µg/L).

These spiking solutions are added to the urine samples to be investigated (U0, U1, U2). For this purpose 1 mL of the appropriate spiking solution is added to 1 mL of each of the urine samples in the autosampler vessels in accordance with the pipetting scheme (Table 2), and the contents are thoroughly mixed. These sample solutions, which have to be freshly prepared for each urine sample, are diluted in a ratio of 1:10 with 1 M nitric acid, and then they are subjected to further sample processing.

Table 2. Pipetting scheme to prepare the spiked urine sample solutions for the standard addition procedure and the resulting concentrations in µg/L

Designation	Volume of the spiking solutions			Volume of Urine (1:10) [mL]	Final volume [mL]	Spiked concentration [µg/L]
	0 [mL]	A [mL]	B [mL]			
U0	1	–	–	1	2	+ 0
U1	–	1	–	1	2	+ 2
U2	–	–	1	1	2	+ 4

3 Specimen collection and sample preparation

3.1 Specimen collection and storage

As is the case for all trace element analyses, it is essential to ensure that the reagents are of the highest possible purity and that the vessels are thoroughly clean. This also applies to sample collection.

EDTA Monovettes (e.g. from Sarstedt) are used to take blood specimens. The blood samples must be centrifuged to separate the serum as soon as possible after collection. The serum is transferred to 10 mL polyethylene tubes with stoppers.

On principle, the serum samples must be stored in the deep-freezer at approximately $-18\,°C$ until analysis. After thawing (at room temperature), they must be mixed thoroughly.

To prevent any possible exogenous contamination, each of the plastic vessels for sample collection must be cleaned before use by leaving them filled with 1 M nitric acid for at least 2 hours, rinsing them thoroughly with ultrapure water and drying them.

The urine should be collected and stored in polyethylene vessels. If the determination cannot be carried out immediately, the urine can be stored in the refrigerator for about 1 week at approx. $+4\,°C$. For longer storage it is advisable to keep the samples in the deep-freezer (at approx. $-18\,°C$).

The urine samples are thawed and brought to room temperature for further processing.

4 Operational parameters

The temperature program for graphite furnace AAS (GFAAS) must be optimally adjusted to suit the relevant conditions (drying, thermal pre-treatment to separate the matrix, vaporisation). The sample gas flow (nebulizer gas) must also be adjusted in accordance with the length of the tube connection to the ICP.

The following instrumental conditions and settings were optimised for the combination of devices used in this case (ICP-MS ELAN-5000, ETV HGA 600, AS-60 autosampler, from Perkin-Elmer). These settings serve only as guidelines. The individual parameters must be adjusted for the actual instruments used. It may be necessary to optimise additional settings or parameters on the instruments of other manufacturers.

4.1 Plasma settings

Power supply: 1.2 kW
CEM voltage: 4.1 kV
Sample introduction: Electrothermal vaporisation (ETV)
Plasma conditions: Auxiliary gas Ar 15 L/min
                      Nebulizer gas Ar approx. 0.9 L/min (optimisation necessary)
                      Plasma gas Ar 0.8 L/min

Molybdenum

4.2 Mass spectrometric parameters

Sweeps/reading:	1
Readings/replicate:	120
Number of replicates:	2
Points across peak:	1
Resolution:	Normal
Scanning mode:	Peak hop transient
Baseline time:	600 ms
Transfer frequency:	Replicate
Element masses:	95 and 98 amu
Replicate time:	1200 ms
Dwell time:	10 ms

Elemental equations (equations for mathematical correction)

$^{95}Mo = {}^{95}Mo$
$^{98}Mo = {}^{98}Mo - 0.1095 \cdot {}^{101}Ru$

As Ru is present in plasma and urine at levels in the lower ng/L range, this equation is of no importance for correction. The two Mo isotopes are only measured as a control, the quantification is achieved by measuring ^{95}Mo.

4.3 Parameters for electrothermal vaporisation (ETV)

Sample volume:	20 µL
Modifier:	Pd 100 mg/L, Mg 200 mg/L
Modifier volume:	20 µL
Injection temperature:	20 °C
Injection speed:	50%
Read delay:	2 s
Closure delay:	−6 s

Table 3. ETV parameters (I): Temperature program

Program step	Cell temperature [°C]	Time ramp [s]	Time hold [s]	Internal flow [mL]	Gas norm	Type alt	Gas vent	To ICP	Read
1	90	5.0	10	300	x		X		
2	120	5.0	25	300	x		X		
3	700	10.0	20	300	x		X		
4	1500	5.0	10	300	x		X		
5	1500	1.0	5	100	x		X		
6	2650	0.2	5	100	x			x	x
7	2700	1.0	5	300	x			x	
8	20	1.0	20	300	x			x	

Table 4. ETV parameters (II)

Pipette seq.	Pipette solution			Activate from	Cell to	Wash	Repeat from	Sequence	
	Mod. 1	Mod. 2	Sample					to	times
1			0	0	2	0	0	0	
2	x		0	0	0	0	0	0	
3			1	2	2	0	0	0	

5 Analytical determination

At least 100 µL of the freshly spiked sample solutions (plasma or urine) are pipetted into the autosampler vessels. This is sufficient material to carry out 3 replicate measurements (as is usual in ICP-MS) under the conditions described here. The mean value and the standard deviation for each group of replicate measurements are calculated from these values.

A quality control sample is analysed with each analytical series (see Section 7). If any reagent blank values occur, they must be taken into account.

The quantification is carried out using ^{95}Mo, and ^{98}Mo is also analysed as a control.

6 Calibration and calculation of the analytical result

Calibration is performed by means of the standard addition procedure. The 3 solutions of each sample are introduced into the plasma by means of electrothermal vaporisation. A calibration function is obtained for each sample by plotting the mean signal areas of the 3 sample solutions versus the spiked concentrations. The analytical result can then be read off as the negative portion of the concentration axis.

The samples are evaluated with the standard addition procedure, as there are considerable differences between the gradient of the aqueous calibration function and the standard addition function. See also Figure 1.

7 Standardisation and quality control

A quality control standard to check the precision of the method must be included in each analytical series as stipulated in the guidelines of the Bundesärztekammer (German Medical Association) [46, 47] and in the special preliminary remarks to this series [48, 49].

As no reference material for molybdenum is commercially available, it must be prepared in the laboratory. For this purpose, urine or plasma is spiked with a defined quantity of molybdenum. Aliquots of this solution can be stored in the deep-freezer for up to a year and used for quality control. The concentration of this control material should lie in the middle of the most frequently occurring concentration range. The theoretical value and the tolerance range for this quality control material are determined in the course of a pre-analytical period (one analysis of the control material on each of 20 different days) [50].

External quality control can be achieved by participation in round-robin experiments. However, to the best of our knowledge the round-robin experiments for occupational and environmental toxicological analysis carried out in Germany do not take molybdenum into account in their external quality control programme at present [51].

The author investigated 70 plasma samples from persons who were not occupationally exposed to molybdenum. The resulting values for plasma (0.58 ± 0.34 µg/L) showed a very good correlation with the values given in the literature which were approx. 0.5–0.8 µg/L (see also the introductory section). A frequency distribution of these 70 measured plasma samples is shown in Figure 2.

8 Evaluation of the method

8.1 Precision

Pooled plasma from persons who were not exposed to molybdenum at the workplace were determined for a series of 10 standard additions to ascertain the precision in the series.

Pooled urine was also measured in a series of 10 independent standard additions to determine the relevant precision for the urine matrix. The resulting standard deviations and prognostic ranges are documented in Table 5. The Mo concentration of the pooled plasma was calculated as 0.6 µg/L, and the concentration in pooled urine was 12 µg/L.

Furthermore, the same pooled urine (at both concentrations) was processed and analysed on 10 different days. The precision from day to day is also given in Table 5.

Table 5. Precision

	Calculated concentration [µg/L]	Standard deviation (rel.) [%]	Prognostic range [%]
Precision in the series ($n=10$)			
Urine	12	2.0	4.5
Plasma	0.6	4.5	10.0
Precision from day to day:			
Urine	12	3.0	6.7
Plasma	0.6	5.0	11.1

8.2 Accuracy

As no certified reference material for molybdenum is commercially available at present, recovery experiments to check the accuracy were carried out. For this purpose, pooled plasma with a previously determined content of 0.63 µg/L Mo was spiked with 1 µg/L Mo. This plasma was analysed 10 times. The relative recovery was between 97 and 103%. The same procedure was carried out in the case of urine. Pooled urine with a previously determined content of 12 µg/L Mo was spiked with 10 µg/L Mo and analysed 10 times. The resulting relative recovery rate was 98–104%.

8.3 Detection limits

Under the given conditions the detection limits, calculated as three times the ratio of the signal/background noise in the temporal vicinity of the analyte signal, was 0.05 µg/L for the masses 95 and 98. This applies to both the plasma matrix and the urine matrix.

8.4 Sources of error

It is possible to analyse molybdenum in urine and plasma without interference under the given conditions.

An ETV system is absolutely essential to enable reliable determination using a quadrupole ICP-MS. Interference to the direct determination of molybdenum in urine or plasma, i.e. without ETV, is known to be caused by "polyatomic ions" such as BrO^+ and $ArKO^+$, and erroneously high Mo values are simulated. Mathematical corrections are unreliable in the concentration analyte/interfering substance ratio present in this case. Direct determination is only possible with the use of a sector field ICP-MS instrument at a resolution of approx. 4000.

It is well known that the chemical matrix in the graphite furnace influences the sensitivity of GFAAS and changes the gradient of the calibration graphs. Therefore is it absolutely essential to use the standard addition procedure (Figure 1).

Digestion of the sample is not necessary when ETV is used, as the temperature program has already ensured that the sample is dried and that thermal decomposition to separate the main matrix components from the analyte has taken place before vaporisation, thus avoiding interference at the interface.

9 Discussion of the method

The method described here permits reliable and accurate determination of molybdenum in urine and plasma, even at the concentration range due to environmental exposure. The range of relevance to occupational medicine can be determined at the same time. The reliability criteria are regarded as excellent and the examiners of the method were able to duplicate them at their first attempt.

The advantages of coupling ETV and a quadrupole ICP-MS are as follows:
1) Matrix separation by means of an appropriate temperature program (as in GFAAS), and thus avoidance of interference at the interface. Elevated salt concentrations or higher organic concentrations in the samples pose no problems during processing. For these reasons mineralisation of the sample is not necessary.
2) It is a micro method. A total of only about 200–400 µL plasma is required which is of considerable importance, especially when examining children.
3) Some polyatomic interference can be eliminated by suitable selection of the temperature program. Thus elements that would normally require a higher resolution can be analysed using quadrupole ICP-MS.

The disadvantages are:
1) The method is relatively time-consuming due to the use of the standard addition procedure (approx. 10 min/sample). This is necessary due to the well-known matrix effects in the graphite furnace.
2) It is not possible to use an internal standard, as is usual for ICP-MS, as the thermal behaviour of the analyte and the internal standard are generally very different in the graphite furnace.

The evaluation of ^{95}Mo and ^{98}Mo serves only as a control, ^{95}Mo is used for quantification. Although ^{98}Ru causes interference in the evaluation of ^{98}Mo, this is of no significance for the investigated matrices, as the Ru concentrations in plasma and in urine are only in the lower ng/L range.

The method does not make high demands on the laboratory personnel, as the GFAAS and the quadrupole-ICP-MS are easily and rapidly coupled.

If no sector field ICP-MS is available, this method offers a good alternative for the determination of molybdenum in plasma and urine.

Instruments used:
ICP mass spectrometer ELAN-5000 (from Perkin-Elmer, Germany)
ETV HGA 600 (from Perkin-Elmer, Germany) with AS 60 autosampler (from Perkin-Elmer, Germany).

10 References

[1] *G. K. Davies, R. Jorden, J. Kubota, A. Laitinen, G. Matrone, P. M. Newberne, B. L. O'Dell* and *J. S. Webb:* In: Geochemistry and the Environment, Vol. 1, Chap. IX. National Academy of Sciences, Washington, D.C. (1974)

[2] *E. Lugscheider, R. Eck* and *P. Ettmayer:* Chrom, Molybdän, Wolfram und ihre Legierungen als Hochtemperatur- und Verschleißschutzwerkstoffe. Radex Rundschau (Österreichisch-Amerikanische Magnesit-Aktiengesellschaft, A-9545 Radenthein) 75, No. 1/2-1983, 52–84 (1983)

[3] *J. L. De Maris:* Neue Züricher Zeitung, Beilage: Forschung und Technik No. 116, 55 (19 May 1976)

[4] *T. A. Tsongas, R. R. Meglen, P. A. Walravens* and *W. R. Chappell:* Molybdenum in the diet: an estimate of average daily intake in the United States. Amer. J. Clin. Nutr. 33, 1103 (1980)

[5] *P. W. Winston:* Molybdenum. In: Disorders of Mineral Metabolism, ed. by F. Bronner, J. W. Coburn. Academic Press, New York, 295–315 (1981)

[6] *E. J. Underwood:* Trace Elements in Human and Animal Nutrition, 4th ed. Academic Press, New York, 109–131 (1977)

[7] *World Health Organization Expert Committee:* Trace Elements in Human Nutrition. WHO Tech. Rep. Ser. No. 532, 43 (1973)

[8] *The National Academy of Sciences – National Research Council:* Recommended Dietary Allowances Revised 1980. National Dairy Council, Rosemont (1980)

[9] *M. F. Robinson:* Essential trace elements in New Zealand residents. In: Proceedings of New Zealand Workshop on Trace Elements in New Zealand. University of Otago, Dunedin, 50–60 (1981)

[10] *I. J. T. Davies:* The Clinical Significance of the Essential Biological Metals. Heinemann, London, 81–85 (1972)

[11] *C. F. Mills, I. Bremner, B. W. Young* and *N. T. Davies:* Effects of thioanions of molybdenum and tungsten upon copper metabolism. In: Trace Element Metabolism in Man and Animals (TEMA-4), ed. by J.C. McHowell, J.M. Gawthorne, C.L. White. Australian Academy of Science, Canberra, 549–552 (1981)

[12] *R. W. Engel, N. O. Price* and *R. F. Miller:* Copper, manganese, cobalt, and molybdenum balance in pre-adolescent girls. J. Nutr. 92, 197 (1967)

[13] *M. F. Robinson, J. M. McKenzie, C. D. Thomson* and *A. L. van Ry:* Metabolic balance of zinc, copper, cadmium, iron, molybdenum and selenium in young New Zealand women. Brit. J. Nutr. 30, 195 (1973)

[14] *B. S. Iversen, C. Menne, M. A. White, J. Kristiansen, J. M. Christensen* and *E. Sabbioni:* Inductively coupled plasma mass spectrometric determination of molybdenum in urine from a Danish population. Analyst. 123(1), 81–85 (1998)

[15] *Centers of Disease Control:* National report on human exposure to environmental chemicals March 2001. Department of Health and Human Services, CDC, Atlanta Georgia, USA (2001)

[16] *M. Anke* and *M. Glei:* Molybdenum. In: *H.G. Seiler, A. Siegel* and *H. Siegel (eds.):* Handbook on Metals in Clinical and Analytical Chemistry. Marcel Decker Inc., New York, 495–500 (1994)

[17] *J. Versieck* and *R. Cornelis:* Normal levels of trace elements in human blood plasma or serum. Anal. Chim. Acta 116, 217 (1980)

[18] *J. Versieck* and *R. Cornelis:* Trace Elements in Human Plasma or Plasma. CRC Press, Boca Raton, Florida, USA (1989)

[19] *S. P. Ericson, M. L. McHalsky, B. E. Rabinow, K. G. Kronholm, C. S. Arceo, J. A. Weltzer* and *S. W. Ayd:* Sampling and analysis techniques for monitoring serum for trace elements. Clin. Chem. 32, 1350 (1986)

[20] *K. Kasperek, G. V. Iyengar, J. Kiem, H. Bosberg* and *L. E. Feinendegen:* Elemental composition of platelets. III. Determination of Ag, Au, Cd, Co, Cr, Cs, Mo, Rb, Sb and Se in normal human platelets by neutron activation analysis. Clin. Chem. 25, 711 (1979)

[21] *J. Versieck, J. Hoste, L. Vanballenberghe, F. Barbier, R. Cornelis* and *I. Waelput:* Serum molybdenum in diseases of the liver and biliary system. J. Lab. Clin. Med. 97, 535 (1981)

[22] V. V. Kovalsky, G. A. Yaravaya and D. M. Shamavonyan: Changes of purine metabolism in man and animals under conditions of biogeochemicals molybdenum provinces. Zh. Obshch. Biol. 22, 179 (1961)
[23] P.A. Walravens, R. Moure, C.C. Solomons, W.R. Chappell and G. Bentley: Biochemical changes in workers exposed to molybdenum dusts. In: *E.M. Kirchgessner (ed.):* Trace Element Metabolism in Man and Animals, Vol. III. Inst. für Ernährungsphysiologie der Technischen Universität München, Freising-Weihenstephan, 577–581 (1978)
[24] P.A. Walravens, R. Moure-Eraso, C.C. Solomons, W.R. Chappell and G. Bentley: Biochemical abnormalities in workers exposed to molybdenum dust. Arch. Environ. Health 34, 302 (1979)
[25] *Deutsche Forschungsgemeinschaft:* MAK- und BAT-Werte Liste, 37th issue. Wiley-VCH Weinheim (2001)
[26] E. C. DeRenzo, E. Kaleita, P. Heytler, J. J. Oleson, B. L. Hutchings and J. H. Williams: The nature of xanthine oxidase factor. J. Amer. Chem. Soc. 75, 753 (1953)
[27] D.A. Richert and W.W. Westerfeld: Isolation and identification of the xanthine oxidase factor as molybdenum. J. Biol. Chem. 203, 915 (1953)
[28] H.R. Mahler, B. Mackler, D.E. Green and R.M. Bock: Studies on metalloflavoproteins. III. Aldehyde oxidase: a molybdoflavoprotein. J. Biol. Chem. 210, 465 (1954)
[29] H.J. Cohen, I. Fridovich and K.V. Rajagopalan: Hepatic sulfite oxidase. A functional role for molybdenum. J. Biol. Chem. 246, 374 (1971)
[30] J. L. Johnson: The molybdenum cofactor common to nitrate reductase, xanthine dehydrogenase, and sulfite oxidase. In: Molybdenum and Molybdenum-Containing Enzymes, ed. by M.P. Coughlan. Pergamon, Oxford, 345–383 (1980)
[31] J. L. Johnson, B. E. Hainline and K. V. Rajagopalan: Characterization of the molybdenum cofactor of sulfite oxidase, xanthine oxidase, and nitrate reductase. Identification of a pteridine as a structural component. J. Biol. Chem. 255, 1783 (1980)
[32] J. L. Johnson, W. R. Waud, K. V. Rajagopalan, M. Duran, F.A. Beemer and S. K. Wadman: Inborn errors of molybdenum metabolism: Combined deficiencies of sulfite oxidase and xanthine dehydrogenase in a patient lacking the molybdenum cofactor. Proc. Nat. Acad. Sci. 77, 3715 (1980)
[33] H.J. Cohen, R. T. Drew, J. I. Johnson and K. V. Rajagopalan: Molecular basis of the biological function of molybdenum. The relationship between sulfite oxidase and the acute toxicity of bisulfate and SO_2. Proc. Nat. Acad. Sci. 70, 3655 (1973)
[34] J. L. Johnson, K. V. Rajagopalan and H. L. Cohen: Molecular basis of the biological function of molybdenum. Effect of tungsten on xanthine oxidase and sulfite oxidase in the rat. J. Biol. Chem. 249, 859 (1974)
[35] J. L. Johnson and K. V. Rajagopalan: Tryptic cleavage of rat liver sulfite oxidase. Isolation and characterization of molybdenum and heme domains. J. Biol. Chem. 252, 2017 (1977)
[36] J. Versieck, J. Hoste, F. Barbier, L. Vanballenberghe, J. De Rudder and R. Cornelis: Determination of molybdenum in human plasma by neutron activation analysis. Clin. Chim. Acta 87, 135 (1978)
[37] J. Versieck, L. Vanballenberghe, G. Lemey, F. Barbier, R. Cornelis and J. De Rudder: Determination of molybdenum in plasma. In: Trace Element Analytical Chemistry in Medicine and Biology, ed. by P. Brätter, P. Schramel. De Gruyter Verlag, Berlin, 273–282 (1980)
[38] L. B. Sorensen: Specific uptake of molybdenum by the liver. J. Lab. Clin. Med. 60, 1020 (1962)
[39] L.B. Sorensen and M. Archambault: Visualization of the liver by scanning with Mo^{99} (molybdate) as tracer. J. Lab. Clin. Med. 62, 330 (1963)
[40] A.C. Altray, H. Rudolph and W.R. Smythe: Mineral metabolism in uremia. Kidney Int. 7, Suppl. 2, 85 (1975)
[41] D. Brune, K. Samsahl and P.O. Wester: A comparison between the amounts of As, Au, Br, Cu, Fe, Mo, Se and Zn in normal and uraemic whole blood by means of neutron activation analysis. Clin. Chim. Acta 13, 285 (1966)
[42] J. Danys and M. Kusleikaite: Die Konzentrationsänderung der Spurenelemente im Plasma bei Rheumatismus und anderen inneren Krankheiten. Z. inn. Med. 26, 718 (1971)
[43] L.O. Plantin: Trace elements in cardiovascular diseases. In: Nuclear Activation Techniques in the Life Sciences. International Atomic Energy Agency, Vienna, 321–332 (1979)

[44] *H. Rudolph, A.C. Alfrey* and *W.R. Smythe:* Muscle and Plasma trace element profile in uremia. Trans. Amer. Soc. Artif. Intern. Organs 19, 456 (1973)

[45] *K. Samsahl, D. Brune* and *P.O.Wester:* Simultaneous determination of 30 trace elements in cancerous and non-cancerous human tissue samples by neutron activation analysis. Int. J. appl. Radiat. Isot. 16, 273 (1965)

[46] *Bundesärztekammer:* Qualitätssicherung der quantitativen Bestimmungen im Laboratorium. Neue Richtlinien der Bundesärztekammer. Dt. Ärztebl. 85, A699 – A712 (1988)

[47] *Bundesärztekammer:* Ergänzung der „Richtlinien der Bundesärztekammer zur Qualitätssicherung in medizinischen Laboratorien". Dt. Ärztebl. 91, C159–C161 (1994)

[48] *J. Angerer* and *K.H. Schaller (eds.)* Analysis of Hazardous Substances in Biological Materials. Volume 1–7, Deutsche Forschungsgemeinschaft, Wiley-VCH, Weinheim (1985–2001)

[49] *J. Angerer* and *K.H. Schaller (eds.):* Analysen in biologischem Material, Deutsche Forschungsgemeinschaft. Loose-leaf collection 14th issue. Wiley-VCH, Weinheim (2000)

[50] *J. Angerer, T. Göen* and *G. Lehnert:* Mindestanforderungen an die Qualität von umweltmedizinisch-toxikologischen Analysen. Umweltmed. Forsch. Prax. 3, 307–312 (1998)

[51] *Ringversuch No. 28:* Qualitätsmanagement in der Arbeits- und Umweltmedizin. Projektgruppe Qualitätssicherung, Organisation: Institut für Arbeits-, Sozial- und Umweltmedizin der Universität Erlangen-Nürnberg (2001)

Author: *P. Schramel*
Examiners: *L. Dunemann, J. Begerow, P. Brätter*

Molybdenum

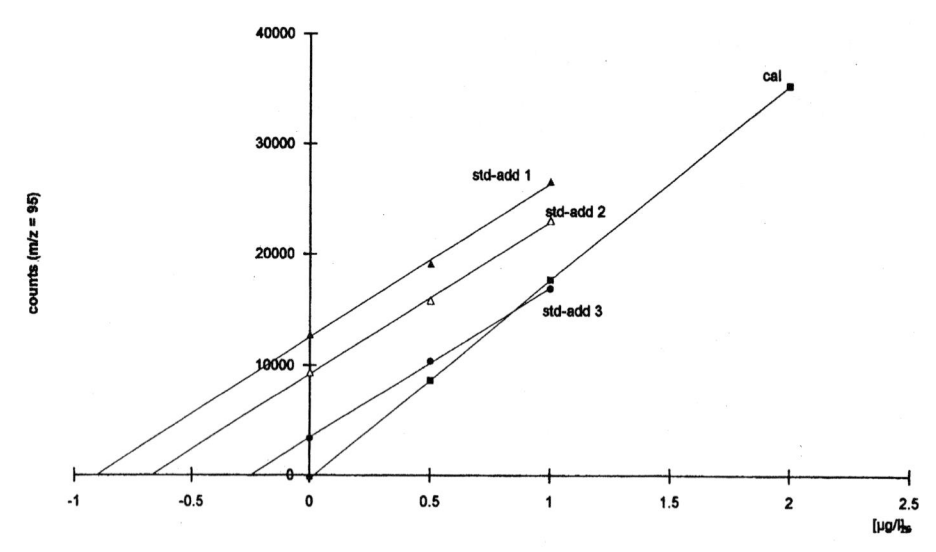

Fig. 1. Comparison of aqueous calibration (cal) and standard addition (std-add 1–3)

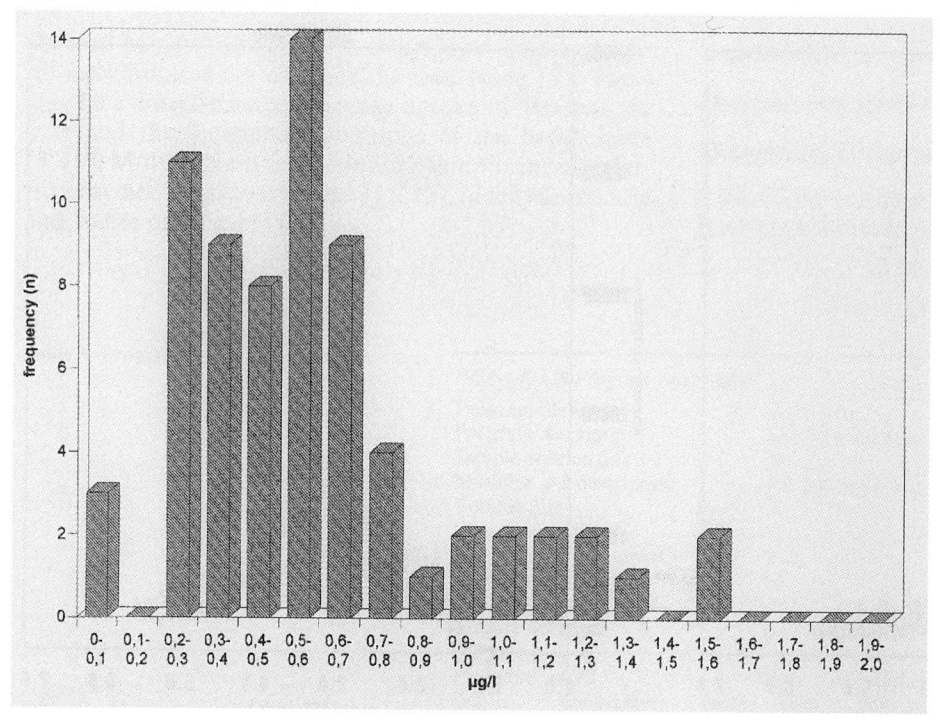

Fig. 2. Frequency distribution of Mo in plasma from 70 samples measured with this method

Organochlorine compounds in whole blood and plasma

Application   Determination in whole blood and plasma

Analytical principle   Capillary gas chromatography/
mass spectrometric detection (MS)

Completed in   May 2001

Summary

The method presented here permits the sensitive and specific quantification of polychlorinated biphenyls, hexachlorobenzene, DDT and DDE, as well as α-, β- and γ-hexachlorocyclohexane. This method can determine the inner exposure of persons who have absorbed these substances from the environment.

Blood and plasma are shaken with formic acid and thus homogenised. The analytes are subsequently extracted in a mixture of hexane and toluene. The extracts are purified and enriched using silica gel, and after capillary gas chromatographic separation they are quantified by means of mass selective detection in the SIM mode. Calibration standards prepared in bovine blood are processed and measured in the same manner as the samples to be analysed. 4,4′-Dibrombiphenyl, δ-hexachlorocyclohexane and Mirex serve as internal standards.

p,p′-DDT

Within-series imprecision: Standard deviation (rel.) s_w = 12.0% or 6.4%
Prognostic range u = 26.7% or 15.8%
at a concentration of 0.25 µg or 1.0 µg p,p′-DDT per litre blood
and where n = 10 or 6 determinations

Between-day imprecision: Standard deviation (rel.) s_w = 14%
Prognostic range u = 35%
at a concentration of 0.4 µg p,p′-DDT per litre blood and where n = 6 determinations

Accuracy: Recovery rate r = 80% at a concentration of 0.5 µg/L

Detection limit: 0.05 µg p,p′-DDT per litre blood

p,p'-DDE

Within-series imprecision: Standard deviation (rel.) s_w = 9.0% or 6.7%
Prognostic range u = 20.1% or 16.6%
at a concentration of 0.25 µg or 1.0 µg p,p'-DDE per litre blood
and where n = 10 or 6 determinations

Between-day imprecision: Standard deviation (rel.) s_w = 7.0%
Prognostic range u = 17.4%
at a concentration of 1.8 µg p,p'-DDE per litre blood and where n = 6 determinations

Accuracy: Recovery rate r = 90% at a concentration of 0.5 µg/L

Detection limit: 0.02 µg p,p'-DDE per litre blood

Hexachlorobenzene (HCB)

Within-series imprecision: Standard deviation (rel.) s_w = 5.3% or 2.7%
Prognostic range u = 11.8% or 6.7%
at a concentration of 0.25 µg or 1.0 µg HCB per litre blood
and where n = 10 or 6 determinations

Between-day imprecision: Standard deviation (rel.) s_w = 8.1%
Prognostic range u = 20.1%
at a concentration of 3.1 µg HCB per litre blood and n = 6 determinations

Accuracy: Recovery rate r = 88% at a concentration of 0.5 µg/L

Detection limit: 0.02 µg HCB per litre blood

α-Hexachlorocyclohexane (α-HCH)

Within-series imprecision: Standard deviation (rel.) s_w = 10.0% or 2.8%
Prognostic range u = 22.3% or 6.9%
at a concentration of 0.25 µg or 1.0 µg α-HCH per litre blood
and where n = 10 or 6 determinations

Between-day imprecision: Standard deviation (rel.) s_w = 7.3%
Prognostic range u = 18.1%

	at a concentration of 0.7 µg α-HCH per litre blood and where $n = 6$ determinations
Accuracy:	Recovery rate $\quad r = 86\%$ at a concentration of 0.5 µg/L
Detection limit:	0.05 µg α-HCH per litre blood

β-Hexachlorocyclohexane (β-HCH)

Within-series imprecision: Standard deviation (rel.) $s_w = 11.0\%$ or 8.1%
Prognostic range $\quad u = 24.5\%$ or 20.1%
at a concentration of 0.25 µg or 1.5 µg β-HCH per litre blood
and where $n = 10$ or 6 determinations

Between-day imprecision: Standard deviation (rel.) $s_w = 8.3\%$
Prognostic range $\quad u = 20.6\%$
at a concentration of 2.2 µg β-HCH per litre blood
and where $n = 6$ determinations

Accuracy: Recovery rate $\quad r = 80\%$ at a concentration of 0.5 µg/L

Detection limit: 0.05 µg β-HCH per litre blood

γ-Hexachlorocyclohexane (γ-HCH)

Within-series imprecision: Standard deviation (rel.) $s_w = 11.0\%$ or 7.4%
Prognostic range $\quad u = 18.3\%$ or 21.8%
at a concentration of 0.25 µg or 1.0 µg γ-HCH per litre blood
and where $n = 10$ or 6 determinations

Between-day imprecision: Standard deviation (rel.) $s_w = 9.0\%$
Prognostic range $\quad u = 22.3\%$
at a concentration of 1.5 µg γ-HCH per litre blood and where $n = 6$ determinations

Accuracy: Recovery rate $\quad r = 85\%$ at a concentration of 0.5 µg/L

Detection limit: 0.05 µg γ-HCH per litre blood

PCB 28

Within-series imprecision: Standard deviation (rel.) s_w = 9.8% or 4.3%
Prognostic range u = 21.8% or 10.7%
at a concentration of 0.25 µg or 1.0 µg PCB 28 per litre blood
and where n = 10 or 6 determinations

Between-day imprecision: Standard deviation (rel.) s_w = 7.2%
Prognostic range u = 17.9%
at a concentration of 3.4 µg PCB 28 per litre blood and where n = 6 determinations

Accuracy: Recovery rate r = 97% at a concentration of 0.5 µg/L

Detection limit: 0.02 µg PCB 28 per litre blood

PCB 52

Within-series imprecision: Standard deviation (rel.) s_w = 9.2% or 4.4%
Prognostic range u = 20.5% or 10.9%
at a concentration of 0.25 µg or 1.0 µg PCB 52 per litre blood
and where n = 10 or 6 determinations

Between-day imprecision: Standard deviation (rel.) s_w = 6.3%
Prognostic range u = 15.6%
at a concentration of 2.4 µg PCB 52 per litre blood and where n = 6 determinations

Accuracy: Recovery rate r = 97% at a concentration of 0.5 µg/L

Detection limit: 0.02 µg PCB 52 per litre blood

PCB 101

Within-series imprecision: Standard deviation (rel.) s_w = 10.1% or 4.7%
Prognostic range u = 22.5% or 11.6%
at a concentration of 0.25 µg or 1.0 µg PCB 101 per litre blood
and where n = 10 or 6 determinations

Between-day imprecision: Standard deviation (rel.) s_w = 5.4%
Prognostic range u = 13.4%

	at a concentration of 2.1 µg PCB 101 per litre blood and where $n = 6$ determinations
Accuracy:	Recovery rate $r = 95\%$ at a concentration of 0.5 µg/L
Detection limit:	0.02 µg PCB 101 per litre blood

PCB 138

Within-series imprecision: Standard deviation (rel.) $s_w = 9.7\%$ or 6.4%
Prognostic range $u = 21.6\%$ or 15.9%
at a concentration of 0.25 µg or 1.0 µg PCB 138 per litre blood
and where $n = 10$ or 6 determinations

Between-day imprecision: Standard deviation (rel.) $s_w = 7.0\%$
Prognostic range $u = 17.4\%$
at a concentration of 2.7 µg PCB 138 per litre blood and where $n = 6$ determinations

Accuracy: Recovery rate $r = 91\%$ at a concentration of 0.5 µg/L

Detection limit: 0.03 µg PCB 138 per litre blood

PCB 153

Within-series imprecision: Standard deviation (rel.) $s_w = 8.4\%$ or 4.9%
Prognostic range $u = 18.7\%$ or 12.1%
at a concentration of 0.25 µg or 1.0 µg PCB 153 per litre blood
and where $n = 10$ or 6 determinations

Between-day imprecision: Standard deviation (rel.) $s_w = 5.9\%$
Prognostic range $u = 14.6\%$
at a concentration of 2.6 µg PCB 153 per litre blood and where $n = 6$ determinations

Accuracy: Recovery rate $r = 93\%$ at a concentration of 0.5 µg/L

Detection limit: 0.03 µg PCB 138 per litre blood

PCB 180

Within-series imprecision: Standard deviation (rel.) s_w = 9.9% or 6.7%
Prognostic range u = 22.1% or 16.6%
at a concentration of 0.25 µg or 1 µg PCB 180 per litre blood
and where n = 10 or 6 determinations

Between-day imprecision: Standard deviation (rel.) s_w = 6.4%
Prognostic range u = 15.8%
at a concentration of 3.2 µg PCB 180 per litre blood and where n = 6 determinations

Accuracy: Recovery rate r = 100% at a concentration of 0.5 µg/L

Detection limit: 0.03 µg PCB 180 per litre blood

Polychlorinated biphenyls (PCB)

All the biphenyls bearing substituted chlorine atoms belong to the group of the polychlorinated biphenyls. A total of 209 so-called congeners are possible. Depending on the degree of chlorination (1 to 10 chlorine atoms), there are 3 monochloro-, 12 dichloro-, 24 trichloro-, 42 tetrachloro-, 46 pentachloro-, 42 hexachloro-, 12 octachloro-, 3 nonachlorobiphenyls and 1 decachlorobiphenyl. The nomenclature of the PCBs was simplified by Ballschmiter and Zell [1] who arranged and numbered them according to the number and position of the chlorine atoms.

The PCB substances were widely used on account of their chemical and physical properties (e.g. inertness, thermal stability, low electrical conductivity) and the fact that they can be relatively simply and inexpensively manufactured. Thus among other applications PCBs are used as hydraulic fluids, industrial fats and oils, heat-exchange media, impregnating agents, non-conductors in the electronics industry, sealants, organic solvents, paper-coating agents, flame-proofing agents for paper, woven materials and wood, and also as plasticisers in paints.

All the commercially available PCB products are composed of a mixture of various chlorinated biphenyls. As a rule, the chlorine content is between 20 and 60%. Information on the chemical composition of the technical mixtures may be derived from their names (e.g. Aroclor 1242, Chlophen A60, etc.). PCB congeners were generally selected at random for analytical determination in scientific studies until the end of the 1980s. In 1988 six congeners (Figure 1) were selected by agreement in Germany as indicators in routine analyses. These particular congeners have been selected largely for practical purposes to minimise the problems of separating all the congeners from each other by chromatography in one analytical run and thus of quantifying them without interference. The concept of indicator congeners [2] is generally accepted and also used worldwide.

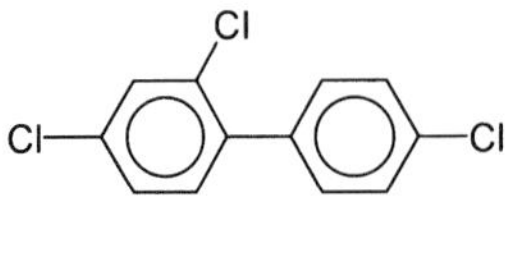

PCB 28 (2,4,4'-Trichlorobiphenyl)

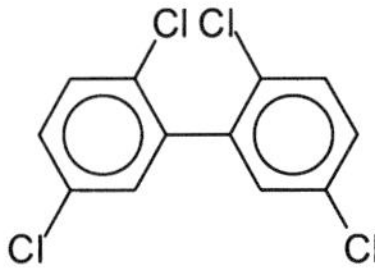

PCB 52 (2,2',5,5'-Tetrachlorobiphenyl)

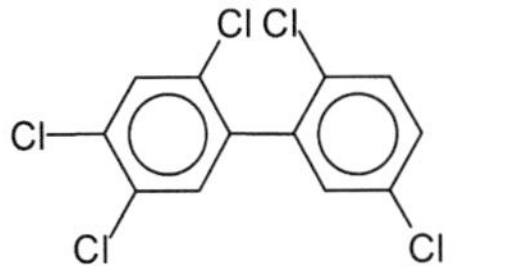

PCB 101 (2,2',4,5,5'-Pentachlorobiphenyl)

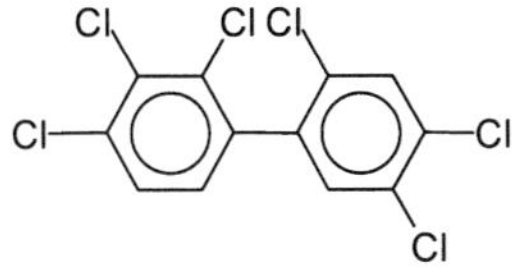

PCB 138 (2,2',3,4,4',5'-Hexachlorobiphenyl)

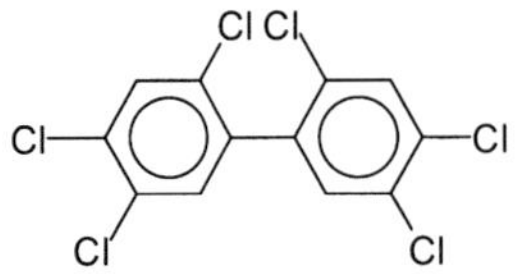

PCB 153 (2,2',4,4'5,5'-Hexachlorobiphenyl)

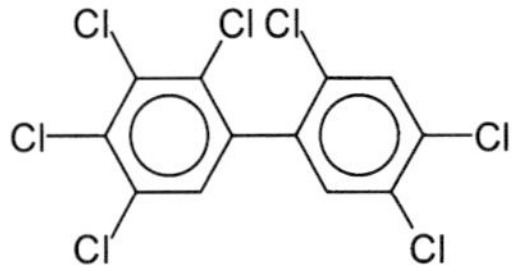

PCB 180 (2,2',3,4,4',5,5'-Heptachlorobiphenyl)

Fig. 1. PCB indicator congeners

Considerable amounts of PCBs are released into the environment due to improper handling (leaking transformers and condensers, disposal in non-sealed refuse sites, accidents, etc.). This causes great problems, as the stable PCBs are not readily decomposed and therefore remain for long periods in the ecosystem where they are not eliminated in many cases but merely relocated. PCBs accumulate in the food chain on account of their lipophilic nature. After the persistence of PCBs in the environment had been recognised, Monsanto and Bayer stopped their production in 1977 and 1983 respectively. The PCB, PCT and VC Ordinance passed in 1989 and the Chemikalien-Verbotsverordnung [Chemical Prohibition Ordinance] that became effective in 1993 ensured that the production, marketing and use of PCBs, polychlorinated terphenyls (PCTs) and vinyl chloride (VC) were forbidden in Germany.

A comprehensive overview of the metabolism of the polychlorinated biphenyls, which has not been completely clarified in detail to date, cannot be presented in this publication. Therefore the reader should refer to the relevant specialised literature (e.g. [3–7]). However, some basic aspects of the absorption, distribution, metabolism and excretion of PCBs in humans are discussed below.

PCBs accumulate in the organism because they are persistent and strongly lipophilic substances. More than 90% of the PCBs found in human tissue have been absorbed with fatty food. The absorption, distribution, metabolism and excretion of PCBs are strongly dependent on the degree of chlorination and the position of the chlorine atoms. Introduction of phenolic OH groups and subsequent conjugation reactions with glucuronic acid or sulphuric acid and glutathione represents the main metabolic pathway in humans, and this leads to accelerated excretion via the kidneys and intestine compared with the non-metabolised PCBs. The low-chlorinated PCBs (e.g. PCB 28, PCB 52, PCB 101) are relatively rapidly metabolised and eliminated, whereas the highly chlorinated congeners (e.g. PCB 138, PCB 153, PCB 180) accumulate in the body.

This also has a considerable effect on the PCB concentrations measured in the blood and fatty tissue of the general population. There is an almost linear relationship between the age of a person and the blood level of the hexachlorinated and heptachlorinated congeners [8], which have mainly been absorbed from the environment over a long period of time. In contrast, if low-chlorinated PCBs (dichlorinated, trichlorinated and tetrachlorinated biphenyls) are detected in the blood, this indicates recent exposure to these substances. Therefore if elevated blood levels of low-chlorinated biphenyls (e.g. PCB 28, PCB 52) compared to the general population are measured, it can be assumed that the exposure to PCBs is current or very recent. Table 1 shows the concentrations of the indicator PCBs found in blood samples from the general population. In general, a distinct drop in the PCB levels in blood samples in Germany has been observed in recent years [8].

Although PCBs generally exhibit a low acute toxicity, chronic exposure to these substances has diverse toxic effects. This applies to the liver in particular. PCBs induce a series of cytochrome P450-dependent enzymes which metabolise extraneous substances. Disorders of the immune systems have been observed in various species of animals after administration of PCBs. In animal studies PCBs are not genotoxic, but they are carcinogenic. The Deutsche Forschungsgemeinschaft's Commission for the Investigation of Health Hazards of Chemical Compounds in the Work Area has assigned technical mixtures of PCBs to category 3B of the carcinogenic working materials [21].

Table 1. Concentrations of polychlorinated biphenyls in blood, plasma or serum samples from persons who have not handled these substances in the course of their work

Substance	n	Group (average age)	Median [ng/L]	Reference
PCB 28	96	Teachers from schools contaminated with PCBs (48)	45 to 98[a]	Gabrio et al. 2000 [9]
	55	Controls (49)	35[a]	
PCB 28	35	General population in Spain (51)	24.5[a,b]	Wingfors et al. 2000 [10]
	26	General population in Sweden (68)	20.5[a,b]	
PCB 52	35	General population in Spain (51)	4.5[a,b]	
	26	General population in Sweden (68)	7[a,b]	
PCB 101	35	General population in Spain (51)	10[a,b]	
	26	General population in Sweden (68)	11.5[a,b]	
PCB 138	35	General population in Spain (51)	1100[a,b]	
	26	General population in Sweden (68)	1150[a,b]	
PCB 153	35	General population in Spain (51)	1500[a,b]	
	26	General population in Sweden (68)	1500[a,b]	
PCB 180	35	General population in Spain (51)	1400[a,b]	
	26	General population in Sweden (68)	1000[a,b]	

Table 1 (continued)

Substance	n	Group (average age)	Median [ng/L]	Reference
PCB 28	120	General population in Sweden (63)	19[b]	Glynn et al. 2000 [11]
PCB 52			<10[b]	
PCB 101			18[b]	
PCB 138			670[b]	
PCB 153			1475[b]	
PCB 180			1030[b]	
PCB 138	13	Immigrants from the former Yugoslavia (27)	200	Schmid et al. 1997 [12]
	29	Immigrants from the former USSR (26)	400	
	28	Immigrants from Asia (28)	<100	
	33	Immigrants from Africa (27)	<100	
	34	General population in Germany (26)	530	
PCB 153	13	Immigrants from the former Yugoslavia (27)	250	
	29	Immigrants from the former USSR (26)	400	
	28	Immigrants from Asia (28)	<100	
	33	Immigrants from Africa (27)	200	
	34	General population in Germany (26)	830	
PCB 180	13	Immigrants from the former Yugoslavia (27)	130	
	29	Immigrants from the former USSR (26)	<100	
	28	Immigrants from Asia (28)	<100	
	33	Immigrants from Africa (27)	<100	
	34	General population in Germany (26)	700	

[a] mean value, [b] calculated as ng/g fat in serum, assuming a constant blood fat content of 0.5%.

PCBs can also cause chloracne, oedema of the eyelids, atrophy of the thymus and the spleen, pathological changes to the kidneys and swelling of the meibomian glands. These effects are mainly observed after exposure to technical PCB mixtures. Therefore the possibility that these symptoms are at least partly caused by polychlorinated dibenzofuran impurities formed in the production process cannot be completely ruled out [14].
See Table 7 for the MAK values that have been assigned to PCB mixtures.

Table 2. Reference values (Germany) for polychlorinated biphenyls in stabilised whole blood and blood plasma in µg/L (Source: Umweltbundesamt [German Federal Environmental Agency] [13])

Age	PCB 138		PCB 153		PCB 180	
	Whole blood	Plasma	Whole blood	Plasma	Whole blood	Plasma
7–10	0.5	–	0.5	–	0.3	–
18–25	0.8	0.8	1.0	1.0	0.7	0.8
26–35	1.0	1.5	1.5	1.9	1.0	1.5
36–45	1.3	2.2	2.0	2.8	1.4	2.2
46–55	1.6	3.0	2.5	3.7	1.9	2.9
55–65	1.8	3.7	3.0	4.6	2.2	3.5

1,2,3,4,5,6-Hexachlorocyclohexane (HCH)

The hexachlorocyclohexane group consists of 8 stereoisomers (Figure 2). Only the a,a,a,e,e,e-isomer (3 neighbouring chlorine atoms in the axial position and the remaining chlorine atoms in the equatorial position), which is also known as γ-HCH or lindane, acts as an insecticide.

While a general prohibition of the use of technical HCH mixtures has been in force in the European Union since 1978, the use of lindane as a biocide is still permitted in agriculture and forestry, as a preservative for wood and textiles (moth protection), in veterinary medicine and for external application in human medicine (e.g. Jacutin a medication against lice). Lindane is also added to the insulation for electric cables to protect them from attack by termites (in Germany only for export).

About nine tonnes of HCH waste isomers are formed as by-products for every tonne of lindane that is manufactured. Until about 1972 some of this waste was simply dumped on open disposal sites in the Federal Republic of Germany. From there mainly α-HCH and β-HCH were dispersed in the environment by the effects of wind and evaporation [15]. As a result of the relatively high persistence of the hexachlorocyclohexanes this still poses an environmental problem today. If it is considered that β-HCH with a half-life of 8 to 10 years and α-HCH and γ-HCH with a half-life of over one year remain in the soil, then it becomes clear that even now this path of introduction still represents a considerable exposure source for animals and plants. As a consequence, food contaminated by HCHs is the main exposure route for man. More than 90% of the HCHs absorbed by humans are taken in with their food.

β-HCH accumulates in the organism on account of its relatively long biological half-

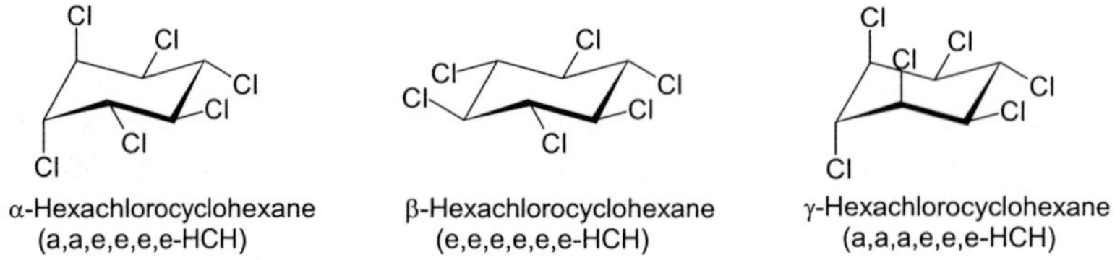

α-Hexachlorocyclohexane (a,a,e,e,e,e-HCH) β-Hexachlorocyclohexane (e,e,e,e,e,e-HCH) γ-Hexachlorocyclohexane (a,a,a,e,e,e-HCH)

Fig. 2. Important HCH isomers

life. This is also reflected in an age-dependency of the blood levels. Table 3 shows the data on the β-HCH exposure of different groups (immigrants shortly after their arrival in Germany), the age-dependent reference values for β-HCH are found in Table 4.

Due to the short half-life in humans ($t_{1/2} = 1$ d) values of less than 0.1 µg/L are generally measured for γ-HCH (lindane) in the general population. As in the case of α-HCH, recent or still existing exposure can be assumed if a level of 0.1 µg per litre blood is exceeded.

Table 3. β-HCH concentrations in plasma samples of the general population

Substance	n	Group (average age)	Median [ng/L]	Reference
β-HCH	13	Immigrants from the former Yugoslavia (27)	<500	Schmid et al. 1997 [12]
	29	Immigrants from the former USSR (26)	1600	
	28	Immigrants from Asia (28)	990	
	33	Immigrants from Africa (27)	<500	
	34	General population in Germany (26)	<500	

Table 4. Reference values (Germany) for β-HCH in whole blood in µg/L (Source: Umweltbundesamt [German Federal Environmental Agency] [19])

Age	β-HCH Whole blood
7–10	0.3
18–25	0.2
26–35	0.4
36–45	0.7
46–55	1.3
55–65	1.3
>65	2.0

The configuration of the HCH isomers determines their metabolism and excretion. β-HCH is metabolised most slowly and is predominantly excreted unchanged with the faeces. In contrast, a-hexachlorocyclohexane and γ-hexachlorocyclohexane are mainly excreted as metabolites via the urine. In man a major part of the a-HCH and γ-HCH is finally converted to various chlorophenols by dehydrochlorination, dechlorination and dehydrogenation [16]. The resulting metabolites are partly eliminated through the kidneys as conjugates of glutathione or glucuronic acid, as sulphate conjugates or as mercapturic acids.

Hexachlorocyclohexanes are not mutagenic. Carcinogenesis experiments with lindane resulted in neoplastic foci in the liver of rats, mice showed an elevated incidence of liver tumours. No irritation was observed in initiation/promotion studies on rats, but a tumour-promoting effect was reported for lindane [17].

a- and β-HCH also exhibited tumour-promoting properties in rats and mice [18].

A MAK value exists for the hexachlorocyclohexanes, lindane has been assigned a MAK and a BAT value (see Table 7).

Hexachlorobenzene (HCB)

Hexachlorobenzene (Figure 3) was mainly used as a pesticide for treating cereal seeds until its use was prohibited in Germany in 1974. In addition, it is formed as a by-product in industrial chemical production, such as the manufacture of trichloroethylene and tetrachloroethylene, during the chlorination of hydrocarbons and in the course of the production of other biocides such as pentachlorophenol or HCH. In the environment HCB mainly accumulates in the soil and water sediments. This is primarily due to the use of hexachlorobenzene as a fungicide for seeds and is also caused by industrial emissions. As a result of its half-life in the soil of 3 to 6 years HCB accumulates in the food chain. Once again fatty foods are the most important source of exposure for humans. However, a continuous reduction in hexachlorobenzene contamination in food has been ascertained since its use was forbidden.

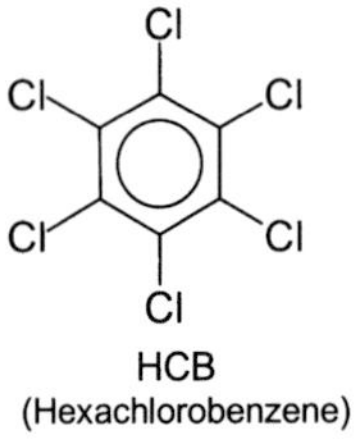

Fig. 3. Hexachlorobenzene

Hexachlorobenzene accumulates in the human body. The corresponding HCB concentrations measured in the blood are again dependent on age. Table 5 shows the age-dependent reference values given by the Umweltbundesamt [German Ministry for the Environment] for Germany.

Table 5. Reference values (Germany) for HCB in whole blood in µg/L (Source: Umweltbundesamt [German Federal Environmental Agency] [19])

Age	HCB Whole blood
7–10	0.4
18–25	0.4
26–35	1.2
36–45	2.1
46–55	2.9
55–65	4.0
>65	4.6

A mass intoxication of humans with fungicides containing HCB over a period of 4 to 5 years led to a high fatality rate. General weakness, skin damage, porphyria, hyposomia, osteoporosis, arthritis and neuritis were described as symptoms.

In animal studies the acute toxicity is low. In rats it is in the order of 3500 to 10,000 mg/kg body weight. The symptoms of intoxication include tremors, convulsions and ataxia. If animals are given HCB orally over a long period of time, they suffer from dermal lesions such as loss of hair, blistering, formation of scabs and hyperpigmentation. Animal studies have shown that the main target organ is the liver. Porphyria is observed, especially in female rats and pigs. Degenerative effects on the liver, pre-neoplastic foci and liver tumours are observed in rats as a result of chronic administration of hexachlorobenzene. Liver tumours are also manifested by mice and hamsters. In addition, tumours of the bile duct, adenomas of the kidneys and phaeochromocytomas of the adrenal gland occur in rats.

The thyroid gland is a further target organ. Hexachlorobenzene also has an immunomodulating influence.

No mutagenic effect of HCB has been observed, but there is evidence of a slight clastogenic effect [20].

HCB has been assigned a BAT value (see Table 7).

p,p'-DDT and p,p'-DDE

The production, import, export, purchase and use of DDT have been prohibited in the Federal Republic of Germany since 1972. DDT is still in use in tropical countries for the purpose of controlling epidemics. DDT has an extraordinarily high persistence and can still be detected in all environmental compartments today as a consequence of its massive worldwide use over many years. DDT is poorly degradable, and on account of its lipophilic character it also accumulates in the human body after intake via the food chain.

DDT is extremely lipophilic and it is readily absorbed via the gastrointestinal tract and the lungs. As a solid DDT is hardly absorbed through the skin, but in solution it can easily penetrate the skin.

The main metabolic pathway of DDT is the enzymatic elimination of HCl to p,p'-dichlorodiphenyldichloroethylene (DDE). DDE is very resistant to further metabolisation and, like DDT, it is also strongly lipophilic. The half-life for the excretion of DDT is approx. 1 year for humans. However, the metabolite DDE is more suitable for detecting long-term exposure to DDT.

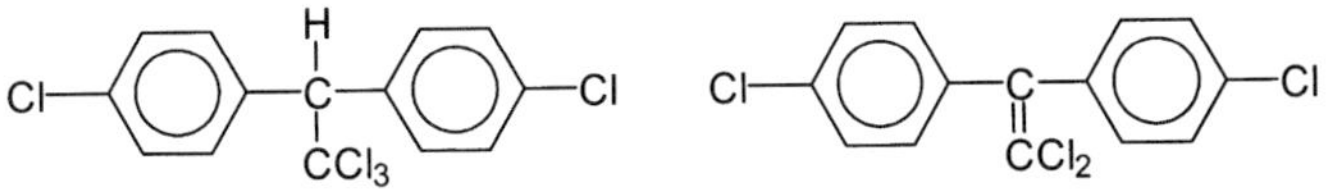

p,p'-DDT
(4,4'-Dichlorodiphenyltrichloroethane)

p,p'-DDE
(4,4'-Dichlorodiphenyldichloroethene)

Fig. 4. DDT and DDE

Table 6 shows DDE concentrations measured in persons with no occupational contact with DDT. In the case of DDE there is also a marked dependence of the blood concentration on age. The mean blood levels in persons over 60 years of age are higher by a factor of 4 to 7 than the levels found in young adults [19].

Table 6. DDE concentrations found in plasma samples of persons who had no occupational contact with DDT

Substance	n	Group (average age)	Median [ng/L]	Reference
DDE	13	Immigrants from the former Yugoslavia (27)	2300	Schmid et al. 1997 [12]
	29	Immigrants from the former USSR (26)	11900	
	28	Immigrants from Asia (28)	16900	
	33	Immigrants from Africa (27)	10850	
	34	General population in Germany (26)	1400	

DDT is a non-systemic contact and intestinal insecticide with a wide application range. DDT is effective against insects by interference with their nerve conduction. The same neurotoxic effect has been observed in humans and other warm-blooded animals.

The lethal oral dose of DDT for humans is approximately 0.1 to 1 g/kg body weight, or in the case of dermal absorption about one order of magnitude higher.

The target organ for chronic DDT intoxication is the liver. The effects of chronic exposure to high doses of DDT range from changes caused to the liver to necrosis of the liver. The NOAEL for man is about 1.5 mg DDT per kg body weight and day.

DDT and its metabolites induce cytochrome P450-dependent enzymes, in particular the P450-B sub-family.

No evidence of a toxic effect on human reproduction has been found for DDT or its metabolites. There is no indication that DDT is mutagenic, except for an impairment of the intercellular communication (fibrobasts of the skin) induced by DDT.

However, numerous animal studies on rats, mice and hamsters have shown that chronic exposure to DDT or DDE leads to liver tumours, lymphomas and tumours of the adrenal cortex.

A MAK value has been assigned to p,p'-DDT (see Table 7).

Table 7. Limit values, H = absorbed through the skin

Substance	Absorbed through the skin	Classification category of carcinogenic substances (DFG [21])	Exposure limit
p,p′-DDT	H	–	MAK: 1 mg/m^3
HCB	H	4	BAT: 150 µg/L plasma
α-HCH, β-HCH mixture	H	–	MAK: 0.5 mg/m^3
γ-HCH (lindane)	H	4	MAK: 0.1 mg/m^3 BAT: 25 µg/L plasma
PCB mixture with 42% chlorine content	H	3B	MAK: 1.1 mg/m^3
PCB mixture with 54% chlorine content	H	3B	MAK: 0.7 mg/m^3

Author: *H.-W. Hoppe, T. Weiss*
Examiners: *M. Ball, J. Lewalter*

Organochlorine compounds in whole blood and plasma

Application   Determination in whole blood and plasma

Analytical principle   Capillary gas chromatography/
mass spectrometric detection (MS)

Completed in   May 2001

Contents

1   General principles
2   Equipment, chemicals and solutions
2.1 Equipment
2.2 Chemicals
2.3 Solutions and conditioning of the clean-up material
2.4 Calibration standards
3   Specimen collection and sample preparation
3.1 Sample preparation
4   Operational parameters
4.1 Operational parameters for gas chromatography and mass spectrometry
5   Analytical determination
6   Calibration
7   Calculation of the analytical result
8   Standardisation and quality control
9   Evaluation of the method
9.1 Precision
9.2 Accuracy
9.3 Detection limits
9.4 Sources of error
10 Discussion of the method
11 References

1 General principles

Blood and plasma are shaken with formic acid and thus homogenised. The analytes are subsequently extracted in a mixture of hexane and toluene. The extracts are purified and enriched using silica gel, and after capillary gas chromatographic separation they are quantified by means of mass selective detection in the SIM mode. Calibration standards prepared in bovine blood are processed and measured in the same manner as the samples to be analysed. 4,4'-Dibrombiphenyl, δ-hexachlorocyclohexane and Mirex serve as internal standards.

2 Equipment, chemicals and solutions

2.1 Equipment

Gas chromatograph with mass selective detector, split-splitless injection system, autosampler and data processing system for evaluation

Capillary gas chromatographic column:
Length: 30 m, inner diameter: 0.25 mm; stationary phase: 5% phenyl / 95% methylpolysiloxane; film thickness: 0.25 µm (e.g. from Hewlett-Packard)

Thermal evaporation block equipped with a fan (e.g. from Barkey)

Laboratory shaker (e.g. Multi-Tube Vortex from Baker)

Laboratory centrifuge (e.g. from Heraeus)

Vacuum centrifuge (e.g. SpeedVac from Savant)

Adjustable 100 µL and 1 mL pipettes (e.g. from Eppendorf)

It is advisable to heat all glassware for 6 hours at 480 °C before use. Alternatively, the glassware can be cleaned by rinsing it several times with acetone.

250 mL and 500 mL glass beakers

100 mL and 500 mL glass cylinders

10 mL and 25 mL volumetric flasks

10 µL to 100 µL glass syringes (e.g. from Hamilton)

Autosampler vials, 2 mL (e.g. from Agilent)

Microvial inserts, 250 µL (e.g. from Agilent)

10 mL Glass tubes with screw caps

20 mL Glass tubes with ground glass stoppers

2.2 Chemicals

Toluene p.a. (e.g. from Merck)

n-Hexane p.a. (e.g. from Merck)

Ethanol p.a. (e.g. from Sigma)

Bovine blood (e.g. from ACILA GMN mbH)

Bovine plasma (e.g. from ACILA GMN mbH)

Formic acid 98–100% (e.g. from Fluka)

Silica gel, 0.063–0.100 mm (Silica Gel 60 from Merck, No. 15101)

Silanised glass wool (e.g. from Supelco)

n-Decane (e.g. from Fluka, No. 30550)

PCB mixture (congeners 28, 52, 101, 138, 153, 180: e.g. from Dr. Ehrenstorfer, L 200301)

Hexachlorocyclohexane isomers (e.g. from Dr. Ehrenstorfer)

p,p'-DDE (e.g. from Dr. Ehrenstorfer)

p,p'-DDT (e.g. from Dr. Ehrenstorfer)

Hexachlorobenzene (HCB) (e.g. from Dr. Ehrenstorfer)

δ-HCH (e.g. from Dr. Ehrenstorfer)

Mirex (e.g. from Dr. Ehrenstorfer)

4,4'-Dibromobiphenyl (e.g. from Aldrich)

2.3 Solutions and conditioning of the clean-up material

The quantities given in this section are sufficient for 20 individual analyses including calibration. The quantities for larger analytical series must be adapted accordingly. All the solutions must be freshly prepared on the day that the analysis is carried out.

Formic acid:
60 mL formic acid are placed in a 250 mL separation funnel. 50 mL n-hexane are added. The mixture is shaken intensively for 5 minutes. After separation of the phases, the lower formic acid phase is poured into a 100 mL glass beaker and used immediately.

Extraction solution (toluene/n-hexane; 1:1, v/v):
210 mL toluene are placed in a 500 ml L glass beaker. 210 mL n-hexane are added while stirring.

Elution solution for cleaning up (toluene/n-hexane; 1:4, v/v):
80 mL toluene are placed in a 500 L glass beaker. 320 mL n-hexane are added while stirring.

Separation column for cleaning up:
A glass column (25 cm, 6 mm diameter, e.g. 5 mL enzyme test pipettes from BRAND, No. 27947) are filled with some glass wool and 1 g silica gel. This column is heated at 480 °C for 6 hours, and after cooling, it is stored in a desiccator until use.

2.4 Calibration standards

2.4.1 Internal standard

Starting solution δ-HCH (internal standard 1):
Approximately 25 mg δ-HCH are weighed exactly into a 25 mL volumetric flask. The volumetric flask is subsequently filled to its nominal volume with toluene (1 g/L).

Starting solution 4,4'-dibromobiphenyl (internal standard 2):
Approximately 25 mg 4,4'-dibromophenyl are weighed exactly into a 25 mL volumetric flask. The flask is subsequently filled to its nominal volume with toluene (1 g/L).

Starting solution Mirex (internal standard 3):
Approximately 25 mg Mirex are weighed exactly into a 25 mL volumetric flask. The flask is subsequently filled to its nominal volume with toluene (1 g/L).

Spiking solution for the internal standard (mixture):
Approx. 2 mL ethanol are placed in a 10 mL volumetric flask. Then 5 µL of the starting solution of δ-HCH in toluene, 5 µL of the starting solution of 4,4'-dibromobiphenyl and 2 µL of the starting solution of Mirex are added using a pipette. The flask is subsequently filled to its nominal volume with ethanol (δ-HCH 0.5 mg/L, 4,4'-dibromobiphenyl 0.5 mg/L; Mirex 0.2 mg/L).

2.4.2 Calibration standards

Starting solution A (PCB mixture):
A commercially available mixture of PCBs Nos. 28, 52, 101, 153 and 180 in iso-octane is used as the starting solution (e.g. Ehrenstorfer, L 200301). The concentration of each PCB congener is 10 mg/L.

Starting solution B (HCH isomers, HCB, p,p'-DDT and p,p'-DDE):
Approximately 25 mg of each substance are weighed exactly into a 25 mL volumetric flask. The flask is subsequently filled to its nominal volume with toluene (1 g/L).

Dilution solution B (HCH isomers, HCB, p,p'-DDT and p,p'-DDE):
Approx. 2 mL toluene are placed in a 10 mL volumetric flask. Then 100 µL of starting solution B are added using a pipette. The flask is subsequently filled to its nominal volume with ethanol (10 mg/L).

Stock solution 1:
Approx. 2 mL ethanol are placed in a 10 mL volumetric flask. 25 µL of dilution solution B and 25 µL of starting solution A are added using a pipette. The flask is then filled to its nominal volume with ethanol (0.025 mg/L).

Stock solution 2:
Approx. 2 mL ethanol are placed in a 10 mL volumetric flask. 250 µL of dilution solution B and 250 µL of starting solution A are added using a pipette. The flask is then filled to its nominal volume with ethanol (0.25 mg/L).

Calibration standards containing between 0.1 and 20 µg/L of the substances are prepared from the stock solutions by dilution with commercially available bovine blood. For this purpose 2.5 mL of the bovine blood or bovine plasma are first placed into a 20 mL glass tube with a ground-glass stopper, and the appropriate volumes of the stock solutions are added using a pipette in accordance with the pipetting scheme shown in Table 8. The sample (calibration standard) is then intensively shaken on a laboratory shaker (Vortex) for 10 seconds.

Table 8. Pipetting scheme for the preparation of calibration standards in blood or plasma

Volume of stock solution 1 [µL]	Volume of stock solution 2 [µL]	Final volume of the calibration standard [mL]	Concentration of the calibration standard [µg/L]
–	–	2.5	0
10	–	2.51	0.1
25	–	2.53	0.25
100	–	2.6	1.0
–	25	2.53	2.5
–	50	2.55	5
–	100	2.6	10
–	200	2.7	20

The given concentrations are based on an ideal final volume of exactly 2.5 mL, which may deviate from the real volume obtained by spiking by no more than 8%. This procedure was selected, as it is relatively difficult to pipette bovine blood. Possible dilution effects were not observed. As evaluation is carried out using internal standards, the deviation is compensated when the calibration curve is plotted and the analytical result is calculated.

As the organochlorine compounds are very stable, the starting solutions, spiking solutions, dilution solutions and stock solutions described in Section 2.4 can be kept in glass vessels in the refrigerator for a practically unlimited period. The actual calibration standards in bovine blood or bovine plasma should be prepared and processed at the same time as the samples.

3 Specimen collection and sample preparation

Approx. 5 to 8 mL blood are withdrawn slowly using a disposable syringe containing an anticoagulant (e.g. EDTA-K Monovettes). If plasma is to be analysed, then it must be prepared by centrifugation (<3500 g) as soon as possible after the blood sample has been taken.
In order to avoid substance loss due to adsorption on the wall of the plastic vessel, the blood or plasma should be transferred to a sealable 10 mL glass tube. The sample can be stored for at least a week at room temperature and at least 21 days in the refrigerator (at approx. 4 °C).

3.1 Sample preparation

2.5 mL blood or plasma, 2.5 mL purified formic acid and 50 µL spiking solution of the internal standards are placed in a 20 mL glass tube with a ground-glass stopper. The sample is homogenised by shaking for 1 minute on the laboratory shaker. Then 10 mL of a mixture of toluene/n-hexane (1:1, v/v) are added using a pipette, and the sample is extracted by shaking on the laboratory shaker. After centrifugation (5 min, 3000 g), 9 mL of the organic phase are taken up with a 10 mL pipette and transferred to a 10 mL glass tube. The samples are evaporated to approx. 100 to 500 µL in a vacuum centrifuge. The concentrated extract is transferred to the prepared clean-up columns using a Pasteur pipette. Then the analytes are eluted by adding 9 mL of a mixture of toluene/n-hexane (1:4, v/v). 50 µL n-decane are added to the eluate as a keeper, and it is evaporated to approx. 100 to 500 µL in a vacuum centrifuge. The solution is subsequently transferred to an autosampler ampoule and concentrated to approx. 50 µL in a stream of nitrogen at 35 °C. The concentrated measurement solution is then transferred into a GC microvial using a pipette.

4 Operational parameters

4.1 Operational parameters for gas chromatography and mass spectrometry

Capillary column:	Material:	Fused silica
	Stationary phase:	DB-5
	Length:	30 m
	Inner diameter:	0.25 mm
	Film thickness:	0.25 µm
Detector:	Mass selective detector (MSD)	

Temperatures:	Column:	Starting temperature 130 °C, 1 minute isothermal, then increase at a rate of 10 °C/min to 295 °C, then 7 min at the final temperature
Injector:	250 °C	
Transfer line:	280 °C	
Carrier gas:	Helium 4.6 at a pre-pressure of 9 psi	
Split:	Splitless, split on after 60 s	
Sample volume:	2 µL	
Evaporation tube:	2 mm inner diameter	
Ionisation type:	Electron impact ionisation (EI)	
Ionisation energy:	70 eV	
Dwell time:	See Table 9	
Electron multiplier:	1400 V + 600 V	

All other parameters must be optimised in accordance with the manufacturer's instructions.

5 Analytical determination

In each case 2 µL is injected into the gas chromatograph for the analytical determination of the blood samples processed as described in Section 3.1. A quality control sample and an aqueous blank sample is analysed with each analytical series. The temporal profiles of the ion fragments shown in Table 9 are recorded in the SIM mode.

Table 9. Retention times, masses and internal standards used for quantification

Compound	Retention time [min]	Masses	Dwell time [ms]	Internal standard
α-HCH	8.3	218.9 216.9*	150	δ-HCH (IS 1)
HCB	8.5	283.9* 285.9	150	δ-HCH (IS 1)
β-HCH	8.9	218.9 216.9*	150	δ-HCH (IS 1)
γ-HCH	9.0	218.9 216.9*	150	δ-HCH (IS 1)
δ-HCH (IS 1)	9.5	218.9 220.9	150	

Table 9 (continued)

Compound	Retention time [min]	Masses	Dwell time [ms]	Internal standard
PCB 28	10.2	255.9* 257.9	70	4,4'-Dibromophenol (IS 2)
PCB 52	11.0	289.9 291.9*	70	4,4'-Dibromophenol (IS 2)
4,4'-Dibromophenol (IS 2)	11.6	312* 310 314	70	
PCB 101	12.7	325.8* 327.8	40	4,4'-Dibromophenol (IS 2)
p,p'-DDE	13.3	316.0 318.0*	40	δ-HCH (IS 1)
PCB 153	14.4	359.8* 361.8	40	Mirex
p,p'-DDT	14.9	235.0* 237.0	40	δ-HCH (IS 1)
PCB 138	15.0	359.8* 361.8	40	Mirex
PCB 180	16.3	393.7* 395.7	80	Mirex
Mirex (IS 3)	16.9	272* 270 274	80	

The masses marked * are used for quantitative evaluation.

The retention times shown in Table 9 serve only as a guide. Users of the method must satisfy themselves of the separation power of the capillary column used and the resulting retention behaviour of the substances. Figure 5 shows an example of a chromatogram of a processed blood standard spiked with 5 µg/L. The chromatogram of a processed human blood sample is shown in Figure 6.

6 Calibration

The calibration standards (Section 2.4.2) are processed in the same manner as the blood samples (Section 3.1) and analysed by gas chromatography/mass spectrometry as described in Sections 4 and 5. Calibration graphs are obtained by plotting the quotients of the peak areas of the analytes and that of the relevant internal standard (see Table 9) as a function of the concentrations used. It is unnecessary to plot a complete calibration graph for every analytical series. It is sufficient to analyse one calibration standard for every analytical series. The ratio of the result obtained for this standard and the result for the equivalent standard in the complete calibration graph is calculated. Using this quotient, each result read off the calibration graph is adjusted for the relevant series.

New calibration graphs should be plotted if the quality control results indicate systematic deviation.
The calibration graph is linear between the detection limit and 20 µg per litre blood.

7 Calculation of the analytical result

Quotients are calculated by dividing the peak areas of the analytes by that of the relevant internal standard. These quotients are used to read off the corresponding concentration of the analytes in µg per litre blood from the relevant calibration graph. If the bovine blood used to prepare the calibration standards exhibits background interference, the resulting calibration graph must be shifted in parallel so that it passes through the zero point of the coordinates. (The concentrations of the background exposure can be read off from the point where the graph intercepts the axis before parallel shifting in each case.) If the aqueous blank solution indicates that there are reagent blank values, the source of the contamination must be identified and eliminated.

8 Standardisation and quality control

Quality control of the analytical results is carried out as stipulated in the guidelines of the Bundesärztekammer (German Medical Association) [22, 23] and in the special preliminary remarks to this series. In order to determine the precision of the method a spiked bovine blood sample containing a constant concentration of the analytes is analysed. As material for quality control is not commercially available, it must be prepared in the laboratory. For this purpose, bovine blood is spiked with a defined quantity of the analytes. A six-month supply of this control material is prepared, divided into aliquots in sealable 10 mL glass tubes and stored in the deep-freezer. The concentration of this control material should lie within the decisive concentration range. The theoretical value and the tolerance range for this quality control material are determined in the course of a pre-analytical period (one analysis of the control material on each of 20 different days) [24–26].
External quality control can be achieved by participation in round-robin experiments. The Deutsche Gesellschaft für Arbeits- und Umweltmedizin (German Association for Occupational and Environmental Medicine) offers polychlorinated biphenyls, hexachlorocyclohexane as well as DDT and DDE as parameters for toxicological occupational and environmental analyses in their round-robin programme [27].

9 Evaluation of the method

9.1 Precision

Bovine blood samples spiked to give concentrations of 0.25 µg/L and 1 µg/L were processed and analysed to check the precision in the series. Ten replicate determinations of these blood samples yielded the precision in the series documented in Table 10.

Table 10. Precision in the series

Parameter	n	Concentration [µg/L]	Standard deviation (rel.) [%]	Prognostic range [%]
p,p'-DDT	10	0.25	12.0	26.7
	6	1	6.4	15.8
p,p'-DDE	10	0.25	9.0	20.1
	6	1	6.7	16.6
HCB	10	0.25	5.3	11.8
	6	1	2.7	6.7
α-HCH	10	0.25	10.0	22.3
	6	1	2.8	6.9
β-HCH	10	0.25	11.0	24.5
	6	1	8.1	20.1
γ-HCH	10	0.25	11.0	24.5
	6	1	7.4	18.3
PCB 28	10	0.25	9.8	21.8
	6	1	4.3	10.7
PCB 52	10	0.25	9.2	20.5
	6	1	4.4	10.9
PCB 101	10	0.25	10.1	22.5
	6	1	4.7	11.6
PCB 138	10	0.25	9.7	21.6
	6	1	6.4	15.9
PCB 153	10	0.25	8.4	18.7
	6	1	4.9	12.1
PCB 180	10	0.25	9.9	22.1
	6	1	6.7	16.6

In addition, the precision from day to day was determined. Plasma samples from round-robin experiments [28] of the Deutsche Gesellschaft für Arbeits- und Umweltmedizin [German Society for Occupational and Environmental Medicine] (18th series of round-robin experiments, sample A containing DDE, HCHs, HCB, PCB) and the Arctic Monitoring and Assessment Programme (round-robin experiments 2001, Round 2, sample W-01-05 including p,p'-DDT) were used. These plasma solutions were processed and analysed on 6 different days. The precision results are also shown in Table 11.

Table 11. Precision from day to day

Parameter	n	Theoretical value [μg/L]	Actual value [μg/L]	Standard deviation (rel.) [%]	Prognostic range [%]
p,p'-DDT	6	0.55	0.4	14	35
p,p'-DDE	6	1.8	2.5	7.0	17.4
HCB	6	3.1	4.1	8.1	20.1
α-HCH	6	0.7	0.4	7.3	18.1
β-HCH	6	2.2	2.6	8.3	20.6
γ-HCH	6	1.5	1.5	9.0	22.3
PCB 28	6	3.4	4	7.2	17.9
PCB 52	6	2.4	2.2	6.3	15.6
PCB 101	6	2.1	2.1	5.4	13.4
PCB 138	6	2.7	2.5	7.0	17.4
PCB 153	6	2.6	2.6	5.9	14.6
PCB 180	6	3.2	3.5	6.4	15.8

9.2 Accuracy

The accuracy of the method was checked by means of recovery experiments using human blood samples. For this purpose, 6 different individual samples with a relatively low background level were selected. These samples were spiked with 0.5 and 2.0 μg/L, then processed and measured 6 times. The mean relative recovery rates can be found in Table 12.

Table 12. Mean relative recovery rates and losses due to processing in spiked human blood samples

Parameter	n	Relative recovery (0.5 μg/L) [%]	Relative recovery (2.0 μg/L) [%]	Losses due to processing [%]
p,p'-DDT	6	80	110	26
p,p'-DDE	6	90	95	19
HCB	6	88	90	16
α-HCH	6	86	85	20
β-HCH	6	80	79	5
γ-HCH	6	85	86	18
PCB 28	6	97	99	9
PCB 52	6	97	98	11
PCB 101	6	95	100	16
PCB 138	6	91	97	17
PCB 153	6	93	95	17
PCB 180	6	100	96	14

Furthermore, the accuracy of the method was objectively evaluated by participation in the 19th series of round-robin experiments of the Deutsche Gesellschaft für Arbeits- und Umweltmedizin [German Society for Occupational and Environmental

Medicine] [29]. As the results in Table 13 show, there is an excellent correlation between the theoretical values of the round-robin experiments and the values obtained using this method.

Table 13. Theoretical values and values obtained by this method during the 19th series of round-robin experiments of the DGAUM

Substance	19th Series of round-robin experiments Sample A (environmental range)		19th Series of round-robin experiments Sample B (environmental range)	
	Theoretical value (tolerance range) [µg/L]	Result [µg/L]	Theoretical value (tolerance range) [µg/L]	Result [µg/L]
p,p'-DDE	1.14 (0.66–1.62)	1.0	2.88 (2.0–3.76)	2.6
HCB	2.13 (1.35–2.92	2.2	4.35 (2.94–5.76)	4.5
α-HCH	0.09 (0.03–0.15)	0.08	0.27 (0.15–0.4)	0.36
β-HCH	1.1 (0.7–1.5)	1.35	2.91 (2.1–3.7)	3.1
γ-HCH	1.15 (0.73–1.58	1.0	2.9 (1.9–3.9)	3.0
PCB 28	0.46 (0.26–0.66)	0.43	1.78 (1.03–2.54)	1.4
PCB 52	0.46 (0.28–0.65)	0.40	1.45 (0.932–1.98)	1.2
PCB 101	0.71 (0.41–1.0)	0.65	2.37 (1.47–3.2)	2.2
PCB 138	1.78 (1.19–2.4)	1.4	3.88 (2.67–5.1)	4.0
PCB 153	0.45 (0.28–0.62)	0.43	4.63 (3.33–5.93)	4.7
PCB 180	1.39 (0.97–1.82)	1.4	4.94 (3.51–6.37)	4.5

9.3 Detection limits

Under the conditions described here the detection limits, calculated as three times the signal/noise ratio of the analytical background interference in the temporal environment of the analyte signals, were between 0.02 and 0.05 µg/L. The detection limits for the parameters determined using this method are shown in Table 14.

Table 14. Detection limits in µg/L

Substance	Detection limits
p,p'-DDT	0.05
p,p'-DDE	0.02
HCB	0.02
α-HCH	0.05
β-HCH	0.05
γ-HCH	0.05
PCB 28	0.02
PCB 52	0.02
PCB 101	0.02
PCB 138	0.03
PCB 153	0.03
PCB 180	0.03

9.4 Sources of error

Interference due to matrix components or exogenous substances was occasionally observed in the case of γ-HCH, and this was attributed to the deteriorating separation capability of ageing separation columns. This interference was clearly recognisable from the peak pattern and the isotope ratio. In such cases the 219 mass fraction can be used for evaluation instead. The separation of β-HCH and γ-HCH gives an indication of performance of the separation column. These isomers must be separated to the baseline. Otherwise a new column must be used.

p,p'-DDT is a critical substance. High sensitivity and acceptable precision can only be achieved with clean, deactivated evaporation tubes and good columns.

In addition, it was checked whether the following substances or substance groups caused interference to the method:

Chlordane	Heptachloroepoxide	Phthalates
Oxychlordane	Aldrin	Phosphoric acid esters
Nonachlor	Dieldrin	Hexachloronaphthalene
Chlorophenols	Quintozene	Dichlorfluanid
Chlorobenzenes	Pentachloroaniline	Tolylfluanid
Methoxychlor	Endosulfan	Chlorthalonil
Methoxyolefin	Musk xylene	Polyaromatic hydrocarbons
Heptachlor	Musk ketone	Pyrethroids
Endrin		

Toxaphene congeners (nomenclature according Parlar): Palar 26, 32, 50 and 63

No interference was observed under the conditions given for the method.

The evaporation of the blood extract and the eluate from the silica gel column can be a critical step. In order to prevent the loss of HCH isomers, PCB 28 and PCB 52, the sample may not be evaporated to dryness. n-Decane, which is added as a keeper to prevent losses, may be contaminated by phthalates, but no analytical problems result from the impurities. Decane from Fluka (No. 30550) exhibits a distinctly higher degree of purity than that from Aldrich (No. D90-1). No contamination of the measurement solution caused by the vacuum centrifuge or the evaporation block was observed.

It is essential to find the source of any reagent blank values detected as a result of the inclusion of an aqueous blank solution. The blank values of the individual reagents must be determined, and if necessary the reagents must be replaced by non-contaminated chemicals. Possible contamination due to the glassware or plastic devices used must also be taken into account.

10 Discussion of the method

The GC-MS procedure presented here permits the sensitive, specific and reliable determination of polychlorinated biphenyls (6 indicator PCBs), hexachlorocyclohexane isomers, DDT and DDE as well as hexachlorobenzene in blood.

It is based on the DFG method according to Schulte, Lewalter and Ellrich [30, 31]. There are two important differences. Firstly, extraction is performed with toluene/hexane instead of n-heptane or isooctane, and secondly detection is carried out by means of MSD instead of ECD. Thus an improved extractability of the HCH isomers and a generally higher specificity and sensitivity of detection is achieved.

On account of its sensitivity the method is suitable for reliably detecting the background levels of the persistent organochlorine compounds in the general population. The described method is extremely specific due to the use of mass spectrometry. It allows a more reliable determination than an electron capture detector, even in the lower concentration range (<0.1 µg/L). The interference due to the analytical background at concentrations below 0.1 µg/L, which occasionally occurs in methods that are quantified by means of electron capture detector [30, 31], was not observed for this method. The blood concentrations of non-persistent organochlorine compounds (a-HCH and γ-HCH, PCB 28 and PCB 52) are seldom detected and, if so, usually only in the range near the detection limit. However, in the case of current exposure (e.g. from sources in internal rooms and due to occupational exposure) concentrations above 0.1 µg/L can also be present in blood.

Despite the relative laborious processing and evaporation steps, this method is easily usable under routine conditions. An experienced technician can certainly process 35 to 40 samples a day.

The reliability criteria of the method are regarded as good for all the analytes. As the round-robin experiments (Section 9) have shown, accurate results have been achieved for all the analytes using this method.

A mixture of n-hexane/toluene has proved very effective for the extraction of the organochlorine compounds including DDT and the HCH isomers. The only drawback is the relatively time-consuming step required to evaporate the extract. If the analysis is limited to HCB and PCB, then heptane or isooctane can be used without problems [30, 31]. In addition to the liquid/liquid extraction, a clean-up on a column with silica gel is prescribed in order to remove the matrix components, especially the blood lipids. Readers are advised to fill the columns with silica gel themselves and then to heat them to glowing. Deactivation of the silica gel before use is not necessary. This type of clean-up column is easier to keep free of contamination than commercially available columns.

Before extraction, δ-HCH, 4,4'-dibromobiphenyl and Mirex are added as internal standards. If calibration is carried out in the matrix as described, addition of these calibration substances is sufficient to obtain accurate and reliable values. Experiments were performed to carry out calibration directly with stock solutions diluted in n-decane in order to simplify the analytical procedure. However, comparison of the calibration graphs in bovine blood with those in n-decane showed that the gradients differed significantly for some substances. Therefore losses due to processing must be assumed (see Table 12). However, the accuracy of the analytical result is not influenced, as these losses are compensated arithmetically because the results are based on the relevant internal standard.

One of the principal advantages of GC-MS analysis is that stable isotopes can be used as internal standards if necessary ("isotope dilution") in order to optimise the accuracy and precision of the results. This technique is increasingly being used for routine analyses. Deuterated and ^{13}C-labelled isotopes are commercially available (e.g. from Promochem) for all the organochlorine compounds included in this method. However, their cost is relatively high and in the experience of the author general use of such isotopes is not necessary in order to meet the quality criteria.

The sensitivity of the method can be enhanced, if necessary. This can be achieved by increasing the size of the sample batch, by evaporating the extracts to approx. 10 µL in special glass flasks and by the use of negative chemical ionisation (NCI) in combination with mass spectrometric detection.

This method describes the determination of relevant organochlorine compounds in blood and plasma. At present the quality criteria for precision and recovery have only been explicitly established for blood.

The efficiency of the method for determination of organochlorine compounds (also in plasma) has been confirmed by the comparison of results with other laboratories and the successful participation of the author in round-robin experiments (Deutsche Gesellschaft für Arbeits- und Umweltmedizin [German Society for Occupational and Environmental Medicine], Arctic Monitoring and Assessment Programme) for many years.

On principle, plasma is preferable to blood as an investigation matrix, as the organochlorine compounds are only present in the erythrocytes in insignificant amounts, so that higher concentrations can be detected when plasma is used. However, when EDTA whole blood is used, one process step can be saved, and this further reduces the risk of contamination. Reference values for the German general population which provide an orientation with respect to *β*-HCH, HCB and PCB are discussed in the toxicological section of this chapter.

Instruments used:
Hewlett-Packard HP5890 gas chromatograph with Hewlett-Packard HP5971A mass selective detector, with split-splitless injection system, Hewlett-Packard HP7673 autosampler and Hewlett-Packard MS HPChemstation data system.

11 References

[1] *K. Ballschmiter* and *M. Zell:* Analysis of polychlorinated biphenyls (PCB) by gas capillary chromatography. Fresenius Z. Anal. Chem. 302, 20–23 (1980)
[2] *H. Beck* and *W. Mathar:* Analysenverfahren zur Bestimmung von ausgewählten PCB-Einzelkomponenten in Lebensmitteln. Bundesgesundhbl. 28, 1–12 (1985)
[3] *L.J. Fischer, R.F. Seegal, P.E. Ganey, I.N. Pessah* and *P.R. Kodavanti:* Symposium overview: toxicity of non-coplanar PCBs. Toxicol. Sci. 41(1), 49–61 (1998)
[4] *R.D. Kimbrough:* Polychlorinated biphenyls (PCBs) and human health: an update. Crit. Rev. Toxicol. 25(2), 133–163 (1995)
[5] *H. Tryphonas:* Immunotoxicity of polychlorinated biphenyls: present status and future considerations. Exp. Clin. Immunogenet. 11(2/3), 149–162 (1994)
[6] *S. Safe:* Toxicology, structure-function relationship, and human and environmental health impacts of polychlorinated biphenyls: progress and problems. Environ. Health Perspect. 100, 259–268 (1993)

[7] *F. Iverson* and *D.L. Grant:* Toxicology of the polychlorinated biphenyls, dibenzofurans and dibenzodioxins. IARC Sci. Publ. 108, 5–29 (1991)

[8] *U. Heudorf* and *J. Angerer:* Aktuelle PCB-Belastung einer Wohnbevölkerung in Deutschland 1998. Umweltmed. Forsch. Prax. 5(3), 137–142 (2000)

[9] *T. Gabrio, I. Piechotowski, T. Wallenhorst, M. Klett, L. Cott, P. Friebel, B. Link* and *M. Schwenk:* PCB-blood levels in teachers, working in PCB-contaminated schools. Chemosphere 40, 1055–1062 (2000)

[10] *H. Wingfors, G. Lindstrom, B. van Bavel, M. Schuhmacher* and *L. Hardell:* Multivariate data evaluation of PCB and dioxin profiles in the general population in Sweden and Spain. Chemosphere 40, 1083–1088 (2000)

[11] *A.W. Glynn, A. Wolk, M. Aune, S. Atuma, S. Zettermark, M. Maehle-Schmid, P.O. Darnerud, W. Becker, B. Vessby* and *H.O. Adami:* Serum concentrations of organochlorines in men: a search for markers of exposure. Sci. Total Environ. 263, 197–208 (2000)

[12] *K. Schmid, P. Lederer, T. Göen, K.H. Schaller, H. Strebl, A. Weber, J. Angerer* and *G. Lehnert:* Internal exposure to hazardous substances of persons from various continents: investigations on exposure to different organochlorine compounds. Int. Arch. Occup. Environ. Health 69, 399–406 (1997)

[13] *Umweltbundesamt:* Referenzwerte für die PCB-Kongenere Nr. 138, 153, 180 und deren Summe im Humanblut. Bundesgesundhbl. 41(9), 416 (1998)

[14] *G. Koss:* Polychlorierte Biphenyle (PCB). In: *H. Marquardt* and *S.G. Schäfer (eds.):* Lehrbuch der Toxikologie. Spektrum Verlag, Heidelberg (1997)

[15] *DFG – Kommission zur Prüfung von Rückständen in Lebensmitteln:* Hexachlorcyclohexan-Kontamination – Ursachen, Situation und Bewertung, Report IX. Verlag Harald Boltd, Boppard, Deutschland (1982)

[16] *J. Angerer, R. Maass* and *R. Heinrich:* Occupational exposure to hexachlorocyclohexane. VI. Metabolism of gamma-hexachlorocyclohexane in man. Int. Arch. Occup. Environ. Health 52(1), 59–67 (1983)

[17] *H. Greim (ed.):* Lindan. Toxikologisch-arbeitsmedizinische Begründung von MAK-Werten, 27th issue. Wiley-VCH, Weinheim (1998)

[18] *D. Henschler (ed.):* alpha-Hexachlorcyclohexan, beta-Hexachlorcyclohexan. Toxikologisch-arbeitsmedizinische Begründung von MAK-Werten, 9th issue. VCH Verlagsgesellschaft, Weinheim (1983)

[19] *Umweltbundesamt:* Statusbericht zur Hintergrundbelastung mit Organochlorverbindungen in Humanblut. Bundesgesundhbl. 42 (5), 446–448 (1999)

[20] *H. Greim (ed.):* Hexachlorbenzol. Toxikologisch-arbeitsmedizinische Begründung von MAK-Werten, 26th issue. Wiley-VCH, Weinheim (1998)

[21] *Deutsche Forschungsgemeinschaft:* MAK- und BAT-Werte-Liste, Report 37. Wiley-VCH, Weinheim (2001)

[22] *Bundesärztekammer:* Qualitätssicherung der quantitativen Bestimmungen im Laboratorium. Neue Richtlinien der Bundesärztekammer. Dt. Ärztebl. 85, A699–A712 (1988)

[23] *Bundesärztekammer:* Ergänzung der „Richtlinien der Bundesärztekammer zur Qualitätssicherung in medizinischen Laboratorien". Dt. Ärztebl. 91, C159–C161 (1994)

[24] *J. Angerer, T. Göen* and *G. Lehnert:* Mindestanforderungen an die Qualität von umweltmedizinisch-toxikologischen Analysen. Umweltmed. Forsch. Prax. 3, 307–312 (1998)

[25] *G. Lehnert, J. Angerer* and *K.H. Schaller:* Statusbericht über die externe Qualitätssicherung arbeits- und umweltmedizinisch-toxikologischer Analysen in biologischen Materialien. Arbeitsmed. Sozialmed. Umweltmed. 33(1), 21–26 (1998)

[26] *J. Angerer* and *G. Lehnert:* Anforderungen an arbeitsmedizinisch-toxikologische Analysen – Stand der Technik. Dt. Ärztebl. 37, C1753–C1760 (1997)

[27] *Ringversuch Nr. 28.* Qualitätsmanagement in der Arbeits- und Umweltmedizin, Projektgruppe Qualitätssicherung. Organisation: Institut für Arbeits-, Sozial- und Umweltmedizin der Universität Erlangen-Nürnberg (2001)

[28] *Ringversuch Nr. 18.* Qualitätsmanagement in der Arbeits- und Umweltmedizin, Projektgruppe Qualitätssicherung, Organisation: Institut für Arbeits-, Sozial- und Umweltmedizin der Universität Erlangen-Nürnberg (1996)

[29] *Ringversuch Nr. 19.* Qualitätsmanagement in der Arbeits- und Umweltmedizin, Projektgruppe Qualitätssicherung. Organisation: Institut für Arbeits-, Sozial- und Umweltmedizin der Universität Erlangen-Nürnberg (1997)

[30] *E. Schulte, J. Lewalter* and *D. Ellrich:* Polychlorinated Biphenyls. In: *J. Angerer* and *K.-H. Schaller (eds.):* DFG – Analysis of Hazardous Substances in Biological Materials, Vol. 3. Wiley-VCH, Weinheim (1991)

[31] *E. Schulte, J. Lewalter* and *D. Ellrich:* Polychlorierte Biphenyle. In: *J. Angerer* and *K.-H. Schaller (eds.):* DFG – Analysen in biologischem Material. Loose-leaf collection, 10th issue. Verlag Wiley-VCH, Weinheim (1991)

Author: *H.-W. Hoppe, T. Weiss*

Examiners: *M. Ball, J. Lewalter*

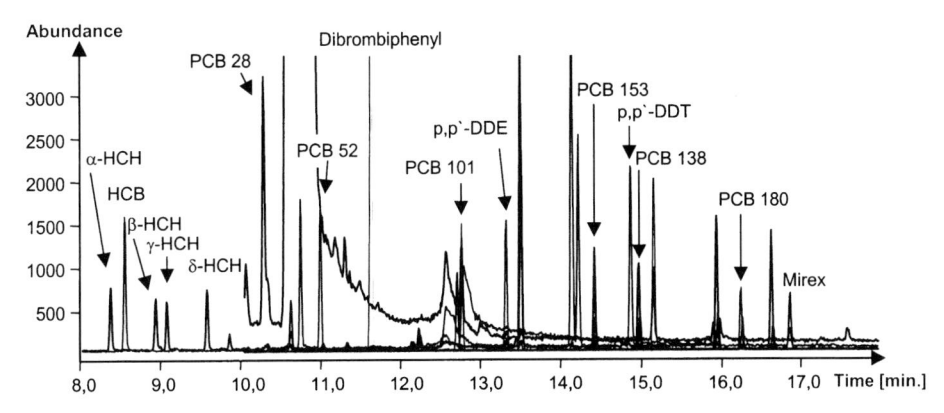

Fig. 5. Example of a chromatogram of a blood sample spiked with 5 µg/L and processed (qualifier ion traces)

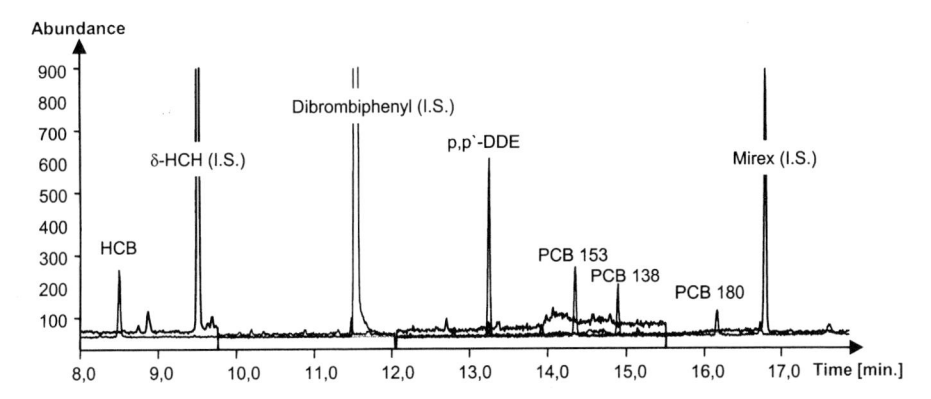

Fig. 6. Example of a chromatogram of a processed individual blood sample of a test person

Oxazaphosphorines: Cyclophosphamide and ifosfamide

Application   Determination in urine

Analytical principle Capillary gas chromatography/
                          mass spectrometric detection (MS)

Completed in   May 2001

Summary

The gas chromatographic procedure with mass selective detection described here enables the determination of the cytostatic oxazaphosphorines, cyclophosphamide and ifosfamide, in the urine of persons who are exposed to these substances at the workplace (e.g. personnel in oncology departments in hospitals). After addition of cyclophosphamide-d_6 as an internal standard in a slightly basic milieu, the urine samples are extracted using methyl-tertiary-butyl ether. After evaporation of the organic phase, the oxazaphosphorines are acetylated with trifluoroacetic acid anhydride. The trifluoroacetyl derivatives are separated from the other components of the urine by means of capillary gas chromatography. The quantitative determination is carried out by mass spectrometry with the help of selected ion monitoring (SIM) of the masses m/z 307.0 as the quantifier and m/z 309.0 as the qualifier. The masses m/z 313.0 (quantifier) and 315.0 (qualifier) are measured for the internal standard. Standard solutions with known cyclophosphamide and ifosfamide concentrations are prepared in urine, then processed and analysed as described in order to obtain calibration functions.

Cyclophosphamide

Within-series imprecision: Standard deviation (rel.) s_w = 4.0 or 3.4%
                          Prognostic range u = 10.0 or 8.5%
                          at a concentration of 3.0 µg or 7.1 µg per litre urine
                          and where n = 6 determinations

Between-day imprecision:	Standard deviation (rel.)	s = 3.2 or 4.1%
	Prognostic range	u = 6.8 or 8.7%
	at a concentration of 3.0 µg or 7.1 µg per litre urine and where n = 15 days	
Accuracy:	Recovery rate	r = 102% or 93%
	at a concentration of 3.0 µg or 7.1 µg per litre urine	
Detection limit:	40 ng per litre urine	

Ifosfamide

Within-series imprecision:	Standard deviation (rel.)	s_w = 10.0 or 10.8%
	Prognostic range	u = 25 or 27%
	at a concentration of 0.5 µg or 1.0 µg per litre urine and where n = 6 determinations	
Between-day imprecision:	Standard deviation (rel.)	s = 13.1 or 13.9%
	Prognostic range	u = 29 or 31%
	at a concentration of 0.5 µg or 1.0 µg per litre urine and where n = 10 days	
Accuracy:	Recovery rate	r = 108% or 101%
	at a concentration of 0.5 µg or 2.0 µg per litre urine	
Detection limit:	50 ng per litre urine	

Oxazaphosphorines

The two oxazaphosphorines, cyclophosphamide [CAS No. 6055-19-2] and ifosfamide [CAS No. 3778-73-2] (Figure 1), are used as cytostatic agents, especially in the chemotherapy of malignant tumours, and also in the case of autoimmune diseases and rheumatic disorders. Both substances are alkylating agents and are known as "prodrugs", i.e. their alkylating effect is induced only after metabolic activation by liver microsomes. In the case of cyclophosphamide the activation proceeds via a hydroxylation to 4-hydroxycyclophosphamide, which forms the actual effective alkylator (N,N-bis-(2-chloroethyl)-phosphoric acid diamide) after cleavage of the nephrotoxic by-product acrolein. Ifosfamide is activated in the liver in a similar manner [1].

Cyclophosphamide: 2-(Bis(2-Chloroethyl)amino)-tetrahydro-2H-1,3,2-oxazaphosphorine-2-oxide

Ifosfamide: 3-(2-Chloroethyl)-2-(2-chloroethylamino)-tetrahydro-2H-1,3,2-oxazaphosphorine-2-oxide

Fig. 1. Structures of cyclophosphamide and ifosfamide

Both substances are used in the kilogram quantity range in clinics and out-patient departments. A survey of 16 German hospitals showed for example that an average of 1.1 kg (maximum 3.0 kg) of cyclophosphamide and 1.2 kg (maximum 4.6 kg) of ifosfamide were administered in 1996.

It is essential to comply with the special safety measures stipulated in TRGS 525 [2] when handling these substances, as alkylating agents can also attack healthy cells. For this reason the IARC [3] has classified cyclophosphamide as carcinogenic in humans (group 1) and ifosfamide as carcinogenic in animal studies (group 2B). Those who are engaged in the manufacture of the drugs (pharmaceutical industry), in preparing the solutions for administration (pharmacy and healthcare personnel) and those who administer the drugs (doctors) in hospitals and out-patient departments run the risk of exposure. The care of patients receiving chemotherapy presents another possibility for exposure through contact with elimination products (e.g. urine, vomit). Even the transport of the cytostatic agents inside and outside the clinic as well as the disposal of waste should not be forgotten. In addition, the entry of these substances into the environment with waste water from the hospital should be taken into account [4], as the biodegradability of some of these substances is limited [5].

To date the only data available on the excretion kinetics of incorporated oxazaphosphorines are from investigations of patients receiving the drugs. Monophasic and biphasic kinetics have been described for both substances with biological half-lives of 1.8 to 9.2 hours in the case of cyclophosphamide [6] and 3.8 to 7.5 hours for Ifosfamide [7, 8]. Renal excretion of the native substances is given as 10 to 50%. The applicability of these data to the low concentrations expected in the case of biomonitoring of occupationally exposed persons must be doubted, as on the one hand very high concentrations are administered during chemotherapy and on the other intravenous administration must be differentiated from intake through the skin or by inhalation. Nevertheless, biomonitoring of the substances in urine within a few days after exposure is a sensible policy. Relevant investigations on the excretion of cyclophosphamide following occupational exposure have already been published [9–12]. On the basis of such findings it is possible to pinpoint improper handling of cytostatic drugs and to verify that any such shortcomings have been eliminated. Air and swab samples are also suitable for primary prevention. These samples can also be analysed with the method presented here.

The expected concentrations in urine samples range from values of the mg/L order excreted by patients to below 1 µg/L for the monitoring of personnel who have contact with cytostatic agents at their workplace. Therefore 1415 urine samples from 100 hospital pharmacy personnel or hospital healthcare workers were investigated by the author of this method using the procedure described here. The results of the study are shown in Table 1.

Table 1. Results of the biological monitoring of persons who had occupational contact with cytostatic drugs

	N	N > DL	N > QL	Median [µg/L]	Maximum [µg/L]
Cyclophosphamide	1415	75	31	0.18	0.95
Ifosfamide	1415	25	12	0.22	2.00

DL = Detection Limit, QL = Quantitation Limit

Authors: *K. Hauff, R. Schierl*
Examiner: *G. Sabbioni*

Oxazaphosphorines: Cyclophosphamide and ifosfamide

Application   Determination in urine

Analytical principle Capillary gas chromatography/
                        mass spectrometric detection (MS)

Completed in   May 2001

Contents

1 General principles
2 Equipment, chemicals and solutions
2.1 Equipment
2.2 Chemicals
2.3 Solutions
2.4 Calibration standards
3 Specimen collection and sample preparation
3.1 Sample preparation
4 Operational parameters
4.1 Operational parameters for gas chromatography and mass spectrometry
5 Analytical determination
6 Calibration
7 Calculation of the analytical result
8 Standardisation and quality control
9 Reliability of the method
9.1 Precision
9.2 Accuracy
9.3 Detection limits
9.4 Sources of error
10 Discussion of the method
11 References

Oxazaphosphorines: Cyclophosphamide and ifosfamide

1 General principles

After addition of cyclophosphamide-d_6 as an internal standard in a slightly basic milieu, the urine samples are extracted using methyl-tertiary-butyl ether. After evaporation of the organic phase, the oxazaphosphorines are acetylated with trifluoroacetic acid anhydride. The trifluoroacetyl derivatives are separated from the other components of the urine by means of capillary gas chromatography. The quantitative determination is carried out by mass spectrometry with the help of selected ion monitoring (SIM) of the masses m/z 307.0 as the quantifier and m/z 309.0 as the qualifier. The masses m/z 313.0 (quantifier) and 315.0 (qualifier) are measured for the internal standard. Standard solutions with known cyclophosphamide and ifosfamide concentrations are prepared in urine, then processed and analysed as described in order to obtain calibration functions.

2 Equipment, chemicals and solutions

2.1 Equipment

Gas chromatograph/mass spectrometer with split injection device and the possibility of selected ion detection and an integration system or data processing system

Precolumn:
Fused silica deactivated by a basic substance, length 5 m, inner diameter 0.32 mm

Gas chromatographic column:
DB XLB (95% dimethyl 5% diphenylpolysiloxane, length 15 m; inner diameter 0.25 mm; film thickness 0.25 µm (e.g. from J&W)

5 µL Syringe for gas chromatography, but the use of an autosampler is preferable

Device for evaporation under a stream of nitrogen

Water bath (e.g. from Gerhardt)

Laboratory shaker (e.g. from Gerhardt)

Rotary evaporator (e.g. from Brand)

100 mL Volumetric flask

50 mL Centrifuge tubes, sealable

10 mL Pointed-bottom tube with screw top

Pipettes: 10–100 µL, 100–1000 µL, 5 mL, 20 mL (e.g. from Eppendorf)

2.5 mL Crimp-cap GC glass vials with 200 µL glass inserts and PTFE-coated crimp caps (e.g. from Agilent)

Tongs for crimping the crimp-cap GC glass vials (e.g. from Agilent)

2.2 Chemicals

Acetonitrile for organic trace analysis (e.g. from Promochem)

Ethyl acetate for organic trace analysis (e.g. from Promochem)

Methyl tert.-butyl ether for organic trace analysis (e.g. from Promochem)

Toluene for organic trace analysis (e.g. from Merck)

Ultrapure water (ASTM type 1) or bidistilled water

Trifluoroacetic acid anhydride ultrapure puris. p.a. (e.g. from Sigma)

Sodium hydrogen carbonate ultrapure puris. p.a. (e.g. from Merck)

Sodium hydroxide p.a. (e.g. from Merck)

Ifosfamide >99% (e.g. Holoxan, from Asta-Medica)

Cyclophosphamide×H_2O >99% (e.g. from Sigma)

Cyclophosphamide-d_6 >99% (e.g. from Phychem)

2.3 Solutions

Sodium hydroxide (0.1 M):
4 g sodium hydroxide are weighed into a 1 L volumetric flask. The flask is subsequently filled to its nominal volume with ultrapure water (4 g/L).

Sodium hydrogen carbonate solution:
100 g sodium hydrogen carbonate are weighed into a 1 L volumetric flask. The flask is subsequently filled to its nominal volume with ultrapure water (100 g/L).

Solutions of the internal standard

Starting solution:
Approximately 10 mg cyclophosphamide-d_6 are weighed exactly into a 10 mL volumetric flask. The volumetric flask is then filled to its nominal volume with acetonitrile. The solution can be stored in the refrigerator for at least 3 months (1 g/L).

Working solution:
100 µL of the starting solution of the internal standard are pipetted into a 10 mL volumetric flask and the flask is filled up to its nominal volume with acetonitrile. This solution must be freshly prepared every week (10 mg/L).

Stock solution:
Each analysis day 100 µL of the working solution of the internal standard are pipetted into a 10 mL volumetric flask and the flask is filled up to its nominal volume with ultrapure water (100 µg/L).

2.4 Calibration standards

Starting solution A: Cyclophosphamide
5 mg cyclophosphamide are weighed exactly into a 10 mL volumetric flask. The volumetric flask is subsequently filled to its nominal volume with ultrapure water. The working solution thus obtained (0.5 g/L) can be kept in the refrigerator for at least one week.

Starting solution B: Ifosfamide
5 mg ifosfamide are weighed exactly into a 10 mL volumetric flask. The volumetric flask is subsequently filled to its nominal volume with ultrapure water. The working solution thus obtained (0.5 g/L) can be kept in the refrigerator for at least one week.

Stock solution A: Cyclophosphamide
100 µL of starting solution A are pipetted into a 100 mL volumetric flask and the flask is filled to its nominal volume with ultrapure water (0.5 mg/L cyclophosphamide).

Stock solution B: Ifosfamide
100 µL of starting solution B are pipetted into a 100 mL volumetric flask and the flask is filled to its nominal volume with ultrapure water (0.5 mg/L ifosfamide).

Calibration standards in concentrations ranging from 0 to 10 µg of the appropriate oxazaphosphorines per litre urine are prepared from these stock solutions by dilution with pooled urine.

Table 2 shows the pipetting procedure to be followed. For this purpose the appropriate quantities of the stock solutions are pipetted into a 100 mL volumetric flask and the flask is filled to its nominal volume with pooled urine.

Table 2. Pipetting scheme for the preparation of the calibration standards

Volume of the stock solutions		Final volume of the calibration standard in urine [mL]	Concentration of the calibration standard [µg/L]
A [µL]	B [µL]		
0	0	100	0
100	100	100	0.5
200	200	100	1.0
400	400	100	2.0
1000	1000	100	5.0
2000	2000	100	10.0

In each case the calibration standards are divided into aliquots of 6 mL in sealable sample vials and stored in the deep-freezer at approx. −18 °C until processing. The calibration standards can be stored in this manner for at least six months.

3 Specimen collection and sample preparation

Urine is collected in sealable polyethylene bottles (which have been cleaned in a dish-washing machine at 80 °C, rinsed twice with distilled water and dried at 70 °C), and it is stored in a deep-freezer at approx. −18 °C until processing. The urine can be stored in this manner for at least six months.

3.1 Sample preparation

The urine is allowed to thaw, whereby any precipitates that have formed are dissolved by warming to 37 °C. After vigorous shaking, 100 µL of the internal standard (stock solution 100 µg/L: cf. Section 2.3) are added to an aliquot of 5 mL urine in a 50 mL centrifuge tube, and the pH value of this mixture is adjusted to about 9 using a pH electrode and dilute sodium hydroxide (if the pH is too high, it must be adjusted by adding 0.1 M HCl). Then extraction is carried out twice in a 50 mL centrifuge tube on a laboratory shaker using 20 mL methyl-tertiary-butyl ether in each case. The samples are centrifuged after each extraction procedure (approx. 2500 g), the organic phase is withdrawn using a 5 mL pipette, and the organic phases are combined in another 50 mL centrifuge tube. In order to purify the organic phase it is shaken with 5 mL of a 10% sodium hydrogen carbonate solution for 5 minutes. After centrifugation the organic phase is withdrawn and then placed in a 50 mL round-bottomed flask, and its volume is reduced to approx. 3 mL on a rotary evaporator (vacuum approx. 430 mbar, water bath 40 °C). The residual organic phase is transferred to a 10 mL pointed-bottom tube and evaporated to dryness in a stream of nitrogen. Then the entire residue is dissolved in 100 µL ethyl acetate and transferred to a 2.5 mL crimp-cap GC vial. After addition of 50 µL trifluoroacetic acid anhydride, the vial is sealed and derivatisation is achieved by heating in the water bath to 70 °C for 30 minutes. After repeating the evaporation of the sample to dryness in a stream of nitrogen, the residue is finally dissolved in 100 µL toluene and transferred using a pipette to a glass insert that fits into the GC vial.

4 Operational parameters

4.1 Operational parameters for gas chromatography and mass spectrometry

Capillary column:
- Material: Fused silica
- Stationary phase: DB XLB
- Length: 15 m
- Inner diameter: 0.25 mm
- Film thickness: 0.25 µm

Precolumn:	Material:	Fused silica, deactivated using a basic substance
	Length:	5 m
	Inner diameter	0.25 mm
Detector:	Mass selective detector (MSD)	
Temperatures:	Column:	1 minute at 100 °C, then increase at a rate of 20 °C/min to 160 °C, then at a rate of 4 °C/min to 200 °C, then at a rate of 20 °C/min to 280 °C, then 5 min at the final temperature
	Injector:	250 °C
	Transfer line:	280 °C
Carrier gas:	Helium with a precolumn pressure of 0.3 bar	
Split:	Splitless, split on after 45 seconds	
Sample volume:	2 µL	
Ionisation type:	Electron impact ionisation (EI)	
Ionisation energy:	70 eV	
Dwell time:	80 ms	

All other parameters must be optimised in accordance with the manufacturer's instructions.

5 Analytical determination

In each case 2 µL of the toluene solution are withdrawn and injected into the gas chromatograph for the analytical determination of the urine samples processed as described in Section 3.1.

If the measured values are above the linear range of the calibration graphs (>10 µg/L), the urine samples are diluted with ultrapure water in the ratio of 1:10, processed and injected anew. A quality control sample is analysed with each analytical series.

The temporal profiles of the ion traces shown in Table 3 are recorded in the SIM mode.

Table 3. Retention times and masses

Compound	Retention time [min]	Masses
Cyclophosphamide	12.84	307*
		309
Ifosfamide	11.68	307*
		309
Cyclophosphamide-d_6 (IS)	12.80	313*
		315

The masses marked with * are used for quantitative evaluation.

The retention times shown in Table 3 serve only as a guide. Users of the method must satisfy themselves of the separation power of the capillary column used and the resulting retention behaviour of the substances. Figure 2 shows examples of chromatograms (m/z 307 and 313) of native urine samples without oxazaphosphorines (A), and with 1.1 µg/L cyclophosphamide (B) and 1.3 µg/L ifosfamide (C).

6 Calibration

The calibration standards (Section 2.4) are processed in the same manner as the urine samples (Section 3.1) and analysed by gas chromatography/mass spectrometry as described in Sections 4 and 5. Calibration graphs are obtained by plotting the quotients of the peak areas of the oxazaphosphorines and that of the internal standard as a function of the concentrations used. It is unnecessary to plot a complete calibration graph for every analytical series. It is sufficient to analyse one calibration standard for every analytical series. The ratio of the result obtained for this standard and the result for the equivalent standard in the complete calibration graph is calculated. Using this quotient, each result read off the calibration graph is adjusted for the relevant series.

The intensity of the qualifier (m/z 309 or m/z 315) is generally about 28–40% of the respective quantifier (m/z 307 or m/z 313). If large deviations are observed, interference can be assumed. New calibration graphs should be plotted if the quality control results indicate systematic deviation. The calibration graph is linear between the detection limits and 10 µg per litre urine (Figure 3).

7 Calculation of the analytical result

Quotients are calculated by dividing the peak areas of the analytes by that of the internal standard. These quotients are used to read off the appropriate concentration of the oxazaphosphorines in µg per litre from the relevant calibration graph. If the urine used to prepare the calibration standards exhibits a background signal, the resulting calibration graph must be shifted in parallel so that it passes through the zero point of the coordinates. (The concentrations of the background signal can be read off from the intersection with the axis in each case.) The results are corrected as described in Section 6.

As a rule it is not to be expected that reagent blank values will have to be taken into account for the oxazaphosphorines.

8 Standardisation and quality control

Quality control of the analytical results is carried out as stipulated in the guidelines of the Bundesärztekammer (German Medical Association) [13, 14] and in the special preliminary remarks to this series. A urine sample containing a constant concentration of the individual oxazaphosphorines is analysed in order to determine the precision of the method. As material for quality control is not commercially available, it must be prepared in the laboratory. For this purpose, pooled urine is spiked with a defined quantity of the individual oxazaphosphorines. The concentration of this control material should lie within the decisive concentration range. Alternatively, urine of patients who are given the relevant oxazaphosphorines as part of their therapy can be used. This urine must be diluted with pooled urine of non-exposed persons to adjust the concentration to the decisive range. A six-month supply of the control material is prepared, divided into aliquots in pierceable ampoules and stored in the deep-freezer. The theoretical value and the tolerance range for this quality control material are determined in the course of a pre-analytical period (one analysis of the control material on each of 20 different days) [15, 16].

9 Reliability of the method

9.1 Precision

The precision in the series was determined with the urine of two patients treated with cyclophosphamide as part of their therapy, and the samples were diluted with pooled urine. These dilute urine samples contained 3.0 and 7.1 µg cyclophosphamide per litre urine. When 6 replicate determinations of these urine samples were carried out, relative standard deviations of between 3.4 and 4.0%, equivalent to prognostic ranges of between 8.5 and 10.0%, were obtained (cf. Table 4). The values for ifosfamide were determined in one patient's urine that was diluted to 0.5 and 1.0 µg/L (Table 4).

Table 4. Precision in the series for the determination of the oxazaphosphorines (n = 6)

Substance	Concentration [µg/L]	Standard deviation (rel.) [%]	Prognostic range [%]
Cyclophosphamide	3.0	4.0	10.0
	7.1	3.4	8.5
Ifosfamide	0.5	10.0	25
	1.0	10.8	27

In addition, the precision from day to day was determined. The same urine samples were used as for the determination of the precision in the series. This urine was processed and analysed on each of 15 and 10 different days. The relative standard deviations were between 3.2 and 4.1% which are equivalent to prognostic ranges of 6.8 and 8.7% (cf. Table 5).

Table 5. Precision from day to day for the determination of the oxazaphosphorines

Substance	Concentration [µg/L]	Standard deviation (rel.) [%]	Prognostic range [%]
Cyclophosphamide (n = 15)	3.0	3.2	6.8
	7.1	4.1	8.7
Ifosfamide (n = 10)	0.5	13.1	29
	1.0	13.9	31

9.2 Accuracy

Recovery experiments were performed to check the accuracy of the method. For this purpose pooled urine without and with addition of defined quantities of cyclophosphamide was analysed 6 times and quantified by reading the value from a calibration graph. The relative recovery rates were 102% for cyclophosphamide at a concentration of 3 µg/L and 93% at a concentration of 7 µg/L. For ifosfamide the recovery rates were 108% at a concentration of 0.5 µg/L and 101% at a concentration of 2.0 µg/L.

Two series of round-robin experiments were carried out in collaboration with other laboratories for external quality assurance. In the first series of round-robin experiments 4 laboratories analysed two concentrations of cyclophosphamide (samples A and B in Table 6), which were obtained by dilution of a patient's urine with urine that did not contain oxazaphosphorines. One laboratory reported results that indicated a systematic error. The measured results found in this laboratory were excluded from the evaluation due to the low recovery (26 and 16%). The 3 remaining participants obtained satisfactory results with respect to the target value. Five laboratories participated in a second series of round-robin experiments (samples C, D and E in Table 6 and samples E, F and G in Table 7). However, only three of them were capable of quantifying ifosfamide. All the results were included in the evaluation. The results show that the detection even of the low cyclophosphamide concentrations was excellent. As expected, the scatter of the values was greater for ifosfamide.

Table 6. Results of two series of round-robin experiments to determine cyclophosphamide in urine

Sample	Participants n	Cyclophosphamide in urine			
		Mean value [µg/L]	SD [%]	Target value [µg/L]	This method
A	4*	3.08	7.5	3.04	3.00
B	4*	6.35	26.6	7.08	6.70
C	5	4.90	17.1	4.10	4.80
D	5	1.12	32.1	1.10	0.97
E	5	1.59	8.2	1.50	1.49

* The results of one participant were evaluated as outliers.

Table 7. Results of the round-robin experiments for the determination of ifosfamide in urine

Sample	Participants n	Ifosfamide in urine [µg/L]			
		Mean value [µg/L]	SD [%]	Target value [µg/L]	This method
E	3	3.54	22.3	3.50	3.30
F	3	1.46	43.8	1.30	1.40
G	3	8.80	39.7	8.50	8.83

As the laboratories participating in the round-robin experiments each used their own extraction and mass spectrometric methods (3 instruments with quadrupole detector, one with ion trap and one high-resolution instrument), the accuracy of this method has been verified.

9.3 Detection limits

Under the conditions for sample preparation and for gas chromatographic-mass spectrometric determination described here, the detection limit was 0.04 µg per litre urine for cyclophosphamide and 0.05 µg per litre urine for ifosfamide. The quantitation limit was about 0.13 µg per litre urine for cyclophosphamide and about 0.24 µg per litre urine for ifosfamide. The detection limit was estimated as 3 times the signal-noise ratio, the quantitation limit as 10 times the signal-noise ratio.

In addition these parameters were calculated from a typical urine calibration graph in accordance with DIN 32645 and are shown in Table 8.

Table 8. Detection and quantitation limits in the urine matrix calculated in accordance with DIN 32 645 or the signal/noise ratio (S/N)

µg/L	Detection limit		Quantitation limit	
	DIN	S/N	DIN	S/N
Cyclophosphamide	0.09	0.04	0.35	0.13
Ifosfamide	0.43	0.05	1.68	0.24

9.4 Sources of error

Occasionally some of the matrix components and non-identified endogenous substances are not completely separated. These substances can accumulate at the beginning of the capillary column and cause interfering peaks in the chromatogram. In addition, on account of their relatively high polarity the trifluoroacetyl derivatives of cyclophosphamide and ifosfamide react with a strong tailing to any decrease in inertness of the GC system. If the separation capability deteriorates, interference to the signals of the target ion traces used for quantification m/z = 307.0 and the qualifier m/z = 309.0 can occur due to partial overlapping. Therefore the injector insert must be cleaned or replaced approximately every 40–60 samples in order to prevent such problems and to ensure that the quality of the analyses remains constant. Moreover, the use of a deactivated pre-column inserted before the separation column has proved advantageous. This pre-column can be shortened by about 30 cm (until 1 m remains) at the same intervals.

The systematic occurrence of interference to the analytical method has not been observed to date.

10 Discussion of the method

The method for the determination of oxazaphosphorines in urine described here permits a reliable assay of the above-mentioned analytes in the trace range. The method is an advance on a formerly developed GC/MS method [17]. The decisive advantage of this procedure is the use of deuterated cyclophosphamide as the internal standard. This permits the simultaneous quantification of cyclophosphamide and ifosfamide in one analytical run, which almost halves the time required for sample preparation. In addition, different extraction yields are taken into account due to the use of the internal standard. This decisively increases the reliability and the robustness of the method. The exacting standards to be met by the GC separation have already been discussed above (see Section 9.4). Considered overall this method is an excellent tool for biomonitoring in the ppb range. More than 1400 urine samples were successfully analysed as part of a BMBF project (01 HP 694) (see Table 1). It remains to be seen if analysis with HPLC/MS-MS, as published recently [18], offers advantages over the procedure presented here.

Instruments used:
Fisons Instruments GC 8060/MD 800

11 References

[1] *N. E. Sladek:* Metabolism of oxazaphosphorines. Pharmacol. Ther. 37, 301–355 (1988)
[2] Technische Regel für Gefahrstoffe 525: „Umgang mit Gefahrstoffen in Einrichtungen der humanmedizinischen Versorgung". In *Weinmann, Th.:* Gefahrstoffverordnung/Part 2. Carl Heymanns Verlag, Köln (1998)
[3] IARC Monographs on the evaluation of the carcinogenic risk of chemicals to man. Suppl. 7, 182 (1987)
[4] *T. Steger-Hartmann, K. Kümmerer* and *A. Hartmann*: Biological Degradation of cyclophosphamide and its occurrence in sewage water. Ecotoxicol. Environ. Saf. 36, 174–179 (1997)
[5] *T. Kiffmeyer, H. J. Götze, M. Jursch* and *U. Lüders:* Trace enrichment, chromatographic separation and biodegradation of cytostatic compounds in surface water. Fresenius J. Anal. Chem. 361, 185–191 (1998)
[6] *I. Jardine, C. Fenselau, M. Appler, M. N. Kan, R. B. Brundrett* and *M. Colvin:* Quantitation by gas chromatography-chemical ionization mass spectrometry of cyclophosphamide, phosphoramide mustard and nornitrogen mustard in the plasma and urine of patients receiving cyclophosphamide therapy. Cancer Res. 38, 408–415 (1978)
[7] *B. M. Bryant, M. Jarman, M. H. Baker, I. E. Smith* and *J. F. Smyth:* Quantification by Gas Chromatography of N,N'-Di-(2-chloroethyl)-phosphorodiamidic acid in the plasma of patients receiving isophosphamide. Cancer Res. 40, 4734–4738 (1980)
[8] *T. Wagner* and *P. Drings:* Pharmacokinetics and bioavailability of oral ifosfamide. Arzneimittelforschung 36, 878–880 (1986)
[9] *A. S. Ensslin, Y. Stoll* and *A. Pethran:* Biological Monitoring of cyclophosphamide and ifosfamide in urine of hospital personnel occupationally exposed to cytostatic drugs. Occup. Environ. Med. 51, 229–233 (1994)
[10] *M. Hirst, S. Tse, D. G. Mills, L. Levin* and *D. F. White:* Occupational exposure to cyclophosphamide. Lancet 1, 186–188 (1984)
[11] *A. Pethran, K. Hauff, H. Hessel* and *C. H. Grimm:* Biological, cytogenetic and ambient monitoring of exposure to antineoplastic drugs. J. Oncol. Pharm. Practice 4, 57 (1998)
[12] *P. J. Sessink, K. A. Boer, A. P. H. Scheefhals, R. B. M. Anzion* and *R. P. Bos:* Occupational exposure to antineoplastic agents at several departments in a hospital. Int. Arch. Occup. Environ. Health 64, 105–112 (1992)
[13] *Bundesärztekammer:* Qualitätssicherung der quantitativen Bestimmungen im Laboratorium. Neue Richtlinien der Bundesärztekammer. Dt. Ärztebl. 85, A699–A712 (1988)
[14] *Bundesärztekammer:* Ergänzung der „Richtlinien der Bundesärztekammer zur Qualitätssicherung in medizinischen Laboratorien". Dt. Ärztebl. 91, C159–C161 (1994)
[15] *G. Lehnert, J. Angerer* and *K. H. Schaller:* Statusbericht über die externe Qualitätssicherung arbeits- und umweltmedizinisch-toxikologischer Analysen in biologischen Materialien. Arbeitsmed. Sozialmed. Umweltmed. 33(1), 21–26 (1998)
[16] *J. Angerer* and *G. Lehnert:* Anforderungen an arbeitsmedizinisch-toxikologische Analysen – Stand der Technik. Dt. Ärztebl. 37, C1753–C1760 (1997)
[17] *P. J. Sessink* and *M. M. Scholtes:* Determination of cyclophosphamide in urine by gas chromatography-mass spectrometry. J. Chromatogr. B 616, 333–337 (1993)
[18] *C. Minoia, R. Turci, C. Sottani, A. Schiavi, L. Perbellini, S. Angeleri, F. Draicchio* and *P. Apostoli:* Application of high performance liquid chromatography/tandem mass spectrometry in the environmental and biological monitoring of health care personnel occupationally exposed to cyclophosphamide and ifosfamide. Rapid Commun. Mass Spectrom. 12 (20), 1485–1493 (1998)

Authors: *K. Hauff, R. Schierl*
Examiner: *G. Sabbioni*

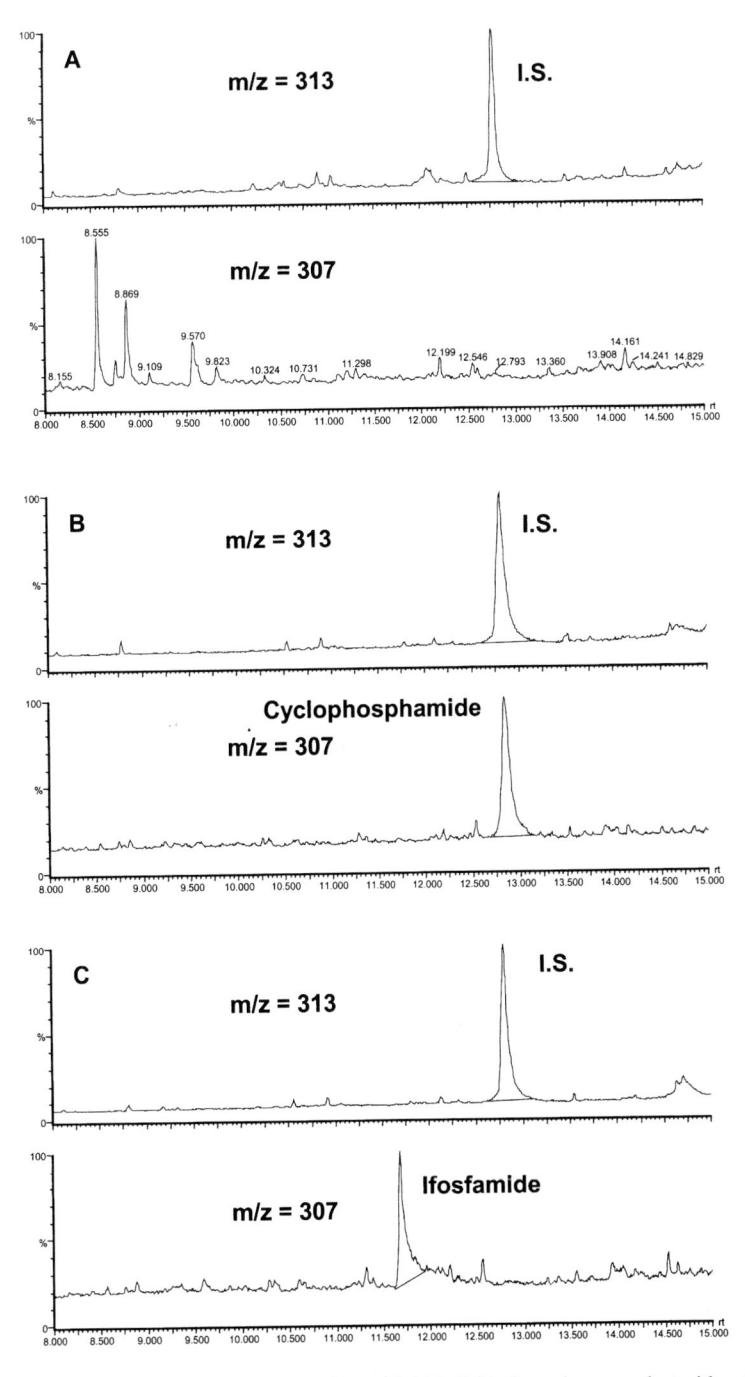

Fig. 2. Chromatograms of the traces m/z 307 and 313 (I.S.) for urine samples without oxazaphosphorines (A), with 1.1 µg/L cyclophosphamide (B) and 1.3 µg/L ifosfamide (C). I.S. = internal standard

Calibration functions for cyclophosphamide and ifosfamide

Cyclophosphamide
y = 0,5264x
r^2 = 0,9997

Ifosfamide
y = 0,1587x
r^2 = 0,9983

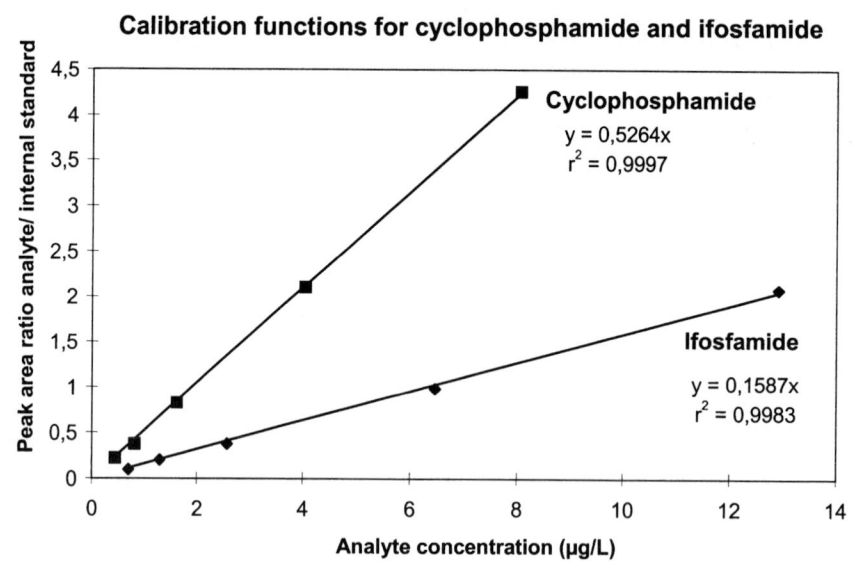

Peak area ratio analyte/ internal standard

Analyte concentration (µg/L)

Fig. 3. Examples of calibration functions in the urine matrix

Members, Guests and ad hoc Experts of the Working Subgroup

Analyses of Hazardous Substances in Biological Materials of the Commission of the Deutsche Forschungsgemeinschaft for the Investigation of Health Hazards of Chemical Compounds in the Work Area

Leader: Prof. Dr. J. Angerer
Institut für Arbeits-, Sozial- und Umweltmedizin
Universität Erlangen-Nürnberg
Schillerstraße 25/29
D-91054 Erlangen

Deputy leader: Dipl.-Ing. K.-H. Schaller
Institut für Arbeits-, Sozial- und Umweltmedizin
Universität Erlangen-Nürnberg
Schillerstraße 25/29
D-91054 Erlangen

Members: Dr. M. Blaszkewicz
Institut für Arbeitsphysiologie
Universität Dortmund
Ardeystraße 67
D-44139 Dortmund

Prof. Dr. H. Greim
Technische Universität München
Institut für Toxikologie und Umwelthygiene
Hohenbachernstraße 15–17
D-85350 Freising-Weihenstephan

Dr. R. Heinrich-Ramm
Zentralinstitut für Arbeitsmedizin
Adolph-Schönfelder-Straße 5
D-22083 Hamburg
verstorben am 12. 02. 2002

Members, Guests and ad hoc Experts of the Working Subgroup

Prof. Dr. A. Kettrup
Institut für Ökologische Chemie
GSF-Forschungszentrum für Umwelt und Gesundheit GmbH
Neuherberg
Ingolstädter Landstraße 1
D-85764 Oberschleißheim

PD Dr. Dr. U. Knecht
Institut und Poliklinik für Arbeits- und Sozialmedizin
Universität Gießen
Aulweg 129/III
D-35392 Gießen

Dr. J. Lewalter
BAYER AG WV-LE
Ärztliche Abteilung: Institut für Biologisches Monitoring
Gebäude L 9
D-51368 Leverkusen

Dr. G. Müller
Institut für Hygiene und Arbeitsmedizin
Universität-GH Essen
Hufelandstraße 55
D-45147 Essen

Dr. M. Müller
Georg-August-Universität Göttingen
Abteilung für Arbeits- und Sozialmedizin
Waldweg 37
D-37073 Göttingen

Prof. Dr. P. Schramel
Institut für Ökologische Chemie
GSF-Forschungszentrum für Umwelt und Gesundheit GmbH
Neuherberg
Ingolstädter Landstraße 1
D-85764 Oberschleißheim

Guests: Dr. F. Alt
Institut für Spektrochemie und angewandte Spektroskopie
Bunsen-Kirchhoff-Straße 11
D-44139 Dortmund

Members, Guests and ad hoc Experts of the Working Subgroup

Dr. M. Bader
Institut und Poliklinik für Arbeitsmedizin
der Medizinischen Hochschule Hannover
Carl-Neuberg-Straße 1
D-30625 Hannover

Dr. M. Ball
Ergo Forschungsgesellschaft GmbH
Geierstraße 1
D-22305 Hamburg

Dr. J. Begerow
Hygieneinstitut des Ruhrgebiets
Institut für Umwelthygiene und Umweltmedizin
Rotthauser Straße 19
D-45879 Gelsenkirchen

Prof. Dr. K.-S. Boos
Institut für Klinische Chemie
Klinikum Großhadern
Marchioninistraße 15
D-81377 München

Prof. Dr. W. Butte
Carl-von-Ossietzky-Universität Oldenburg
FB Chemie
Postfach 2503
D-26129 Oldenburg

Prof. Dr. L. Dunemann
Hygiene-Institut des Ruhrgebiets
Institut für Umwelthygiene und Umweltmedizin
Rotthauser Straße 19
D-45879 Gelsenkirchen

Prof. Dr. H. Emons
Institut für Phytosphäre
Forschungszentrum Jülich GmbH
D-52425 Jülich

Dr. T. Göen
Institut für Arbeitsmedizin
Universitätsklinikum der RWTH Aachen
Pauwelsstraße 30
D-52057 Aachen

Dr. H.-W. Hoppe
Labor Drs. Schiwara/v. Winterfeld
Haferwende 12
D-28357 Bremen

Prof. Dr. E. Richter
Walther-Straub-Institut für Pharmakologie und Toxikologie
Nussbaumstraße 26
D-80336 München

Dr. W. Will
BASF AG
Abt. Arbeitsmedizin und Gesundheitsschutz
D-67056 Ludwigshafen

ad hoc expert: Dr. E. Berger-Preiß
Fraunhofer Gesellschaft ITA
Nikolai-Fuchs-Straße 1
D-30625 Hannover

Prof. Dr. O. Herbarth
UFZ – Umweltforschungszentrum Leipzig-Halle GmbH
Permoserstraße 15
D-04318 Leipzig

Prof. Dr. H.-G. Nolting
Biologische Bundesanstalt für Land- und Forstwirtschaft
Fachgruppe für chemische Mittelprüfung
Messeweg 11/12
D-38104 Braunschweig

Prof. Dr. G. Sabbioni
Walther-Straub-Institut für Pharmakologie und Toxikologie
Nussbaumstraße 26
D-80336 München

Dr. G. Scherer
Labor Prof. Dr. F. Adlkofer
Goethestraße 20
D-80336 München

Dr. J. Wittsiepe
Abt. Hygiene, Sozial- und Umweltmedizin
Ruhruniversität Bochum
Universitätsstraße 150
D-44801 Bochum

Dr. R. Schierl
Institut und Poliklinik für Arbeits- und Umweltmedizin
Ludwig-Maximilians-Universität München
Ziemssenstraße 1
D-80336 München

Scientific secretariat:
T. Weiß
Institut für Arbeits-, Sozial- und Umweltmedizin
Universität Erlangen-Nürnberg
Schillerstraße 25/29
D-91054 Erlangen